LIST OF EFFECTIVE PAGES

DESTROY SUPERSEDED PAGES.

NOTE: The portion of the text affected by the changes is indicated by a vertical line in the outer margins of the page.

TOTAL NUMBER OF PAGES IN THIS PUBLICATION IS 246, CONSISTING OF THE FOLLOWING:

Page	Issue	Page	Issue
* Title	5 Oct 1969	6-2	Original
* A	5 Oct 1969	* 6-3	5 Oct 1969
i	Original	6-4 thru 6-10	Original
ii Blank	Original	* 6-11 thru 6-12	5 Oct 1969
* iii	5 Oct 1969	6-13 thru 6-16	Original
* iv Blank	5 Oct 1969	* 6-17	5 Oct 1969
1-1 thru 1-3	Original	6-18 thru 6-33	Original
* 1-4	5 Oct 1969	6-34 Blank	Original
1-5	Original	7-1 thru 7-21	Original
* 1-6	5 Oct 1969	* 7-22	5 Oct 1969
1-7 thru 1-9	Original	7-23	Original
* 1-10 Blank	Original	* 7-24 thru 7-25	5 Oct 1969
* 2-1 thru 2-20	5 Oct 1969	7-26 thru 7-30	Original
* 3-1	5 Oct 1969	8-1 thru 8-20	Original
3-2 thru 3-10	Original	* 8-21 thru 8-23	5 Oct 1969
* 3-11 thru 3-22	5 Oct 1969	8-24 thru 8-26	Original
* 3-23 Added	5 Oct 1969	9-1 thru 9-13	Original
* 3-24 Blank Added	5 Oct 1969	* 9-14 thru 9-16	5 Oct 1969
4-1 thru 4-3	Original	9-17	Original
* 4-4	5 Oct 1969	9-18 Blank	Original
4-5 thru 4-23	Original	* 10-1 thru 10-7	5 Oct 1969
4-24 Blank	Original	* 10-8 Blank	5 Oct 1969
5-1 thru 5-28	Original	* A-1 thru A-4	5 Oct 1969
* 5-29	5 Oct 1969	* B-1 thru B-3	5 Oct 1969
* 5-30 Blank	5 Oct 1969	* B-4 Blank	5 Oct 1969
* 6-1	5 Oct 1969	* Index 1 thru Index 7	5 Oct 1969
		* Index 8 Blank	5 Oct 1969

* The asterisk indicates pages changed, added, or deleted by the current change.

15 AUGUST 1969

CHANGED 5 OCTOBER 1969

MSFC-MAN-507

SATURN V
FLIGHT
MANUAL

SA 507

THIS PUBLICATION REPLACES MSFC-MAN-507 DATED 15 AUGUST 1969 FOR VEHICLE SA-507; HOWEVER
MSFC-MAN-507 DATED 15 AUGUST 1969 REMAINS THE BASE LINE MANUAL FOR SA-508 AND SA-509.

GEORGE C. MARSHALL   NATIONAL AERONAUTICS AND SPACE ADMINISTRATION

15 AUGUST 1969

CHANGED 5 OCTOBER 1969

Published by Books Express Publishing
Copyright © Books Express, 2012
ISBN 978-1-78039-848-8

Books Express publications are available from all good retail and online booksellers. For
publishing proposals and direct ordering please contact us at: info@books-express.com

TABLE OF CONTENTS

SATURN V FLIGHT MANUAL

SA-507

FOREWORD

This manual was prepared to provide the astronaut with a single source reference as to the characteristics and functions of the SA-507 launch vehicle and AS-507 flight mission. Data on the SA-508 and SA-509 launch vehicles and missions is included, however the 15 August 1969 manual serves as the baseline manual for these two vehicles. A change to the 15 August 1969 manual, incorporating the latest released data, will be released for each vehicle, SA-508 and SA-509, approximately 30 days prior to their respective launch dates.

The manual provides general mission and performance data, emergency detection system information, a description of each stage and the IU, and a general discussion of ground support facilities, equipment, and mission control. A bibliography identifies additional references if a more comprehensive study is desired.

Major hardware associated differences between the Saturn V launch vehicles SA-506 through SA-509 are annotated in the manual. They are identified by reference numbers in the margin adjacent to the new information. These reference numbers refer to footnotes which are located at the end of each section.

This manual is for information only and is not a control document. If a conflict should be discovered between the manual and a control document, the control document will rule.

Recommended changes or corrections to this manual should be forwarded, in writing, to the Saturn V Systems Engineering Management Office (PM-SAT-E) MSFC, Attention: Mr. H. P. Lloyd; or the Crew Safety and Procedures Branch (CF-24), MSC, Attention: Mr. D. K. Warren.

REVISION NOTE

Manual MSFC-MAN-507, dated 15 August 1969, was based on the G mission. The information in this manual describes the vehicle configuration and mission characteristics as defined for the H-1 mission and was prepared from information available approximately thirty days prior to 5 October 1969.

Each page changed to make this revision is identified by a change note at the bottom of the page. Changes of technical significance are identified on these pages by a black bar in the margin opposite the change. Nontechnical changes such as recomposition of pages to accommodate new information, minor rewrite to clarify meaning and correction of typographical errors are not identified by change bars.

TABLE OF CONTENTS

SATURN V SYSTEM DESCRIPTION

The Saturn V system in its broadest scope includes conceptual development, design, manufacture, transportation, assembly, test, and launch. The primary mission of the Saturn V launch vehicle, three-stage-to-escape boost launch of an Apollo Spacecraft, established the basic concept. This mission includes a suborbital start of the third stage (S-IVB) engine for final boost into earth orbit and subsequent reignition to provide sufficient velocity for escape missions including the lunar missions.

LAUNCH VEHICLE DEVELOPMENT

The Saturn launch vehicles are the product of a long evolutionary process stemming from initial studies in 1957 of the Redstone and Jupiter missiles. Early conceptual studies included other proven missiles such as Thor and Titan, and considered payloads ranging from earth orbiting satellites to manned spacecraft such as Dynasoar, Mercury, Gemini, and eventually Apollo.

The Saturn V launch vehicle evolved from the earlier Saturn vehicles as a result of the decision in 1961 to proceed with the Apollo manned lunar mission. As the Apollo mission definition became clear, conceptual design studies were made, considering such parameters as structural dynamics, staging dynamics, and propulsion dynamics.

Design trade-offs were made in certain areas to optimize the launch vehicle design, based on mission requirements. The best combination of design parameters for liquid propellant vehicles resulted in low accelerations and low dynamic loads. Reliability, performance and weight were among primary factors considered in optimizing the design.

Structural design carefully considered the weight factor. Structural rigidity requirements were dictated largely by two general considerations: flight control dynamics and propellant slosh problems. Gross dimensions (diameter & length) were dictated generally by propellant tankage size.

As propulsion requirements were identified, system characteristics emerged: thrust levels, burning times, propellant types and quantities. From these data, engine requirements and characteristics were identified, and the design and development of the total launch vehicle continued, centered around the propulsion systems.

Some of the principal design ground rules developed during the conceptual phase, which were applied in the final design, are discussed in the following paragraphs.

VEHICLE DESIGN GROUND RULES

Safety

Safety criteria are identified by Air Force Eastern Test Range (AFETR) Safety Manual 127-1 and AFETR Regulation 127-9.

Crew safety considerations required the development of an Emergency Detection System (EDS) with equipment located throughout the launch vehicle to detect emergency conditions as they develop. If an emergency condition is detected, this system will either initiate an automatic abort sequence, or display critical data to the flight crew for their analysis and reaction.

Each powered stage is designed with dual redundant range safety equipment which will effect engine cutoff and propellant dispersion in the event of a launch abort after liftoff. Engine cutoff results from closing valves and terminating the flow of fuel and oxidizer. Propellant is dispersed by detonating linear-shaped charges, thereby longitudinally opening the propellant tanks.

Stage Separation

The separation of the launch vehicle stages in flight required design studies involving consideration of many parameters, such as time of separation, vehicle position, vehicle attitude, single or dual plane separation, and the type, quantity, and location of ordnance.

The launch vehicle stages separate in flight by explosively severing a circumferential separation joint and firing retrorocket motors to decelerate the spent stage. Stage separation is initiated when stage thrust decays to a value equal to or less than 10% of rated thrust. A short coast mode is used to allow separation of the spent stage, and to effect ullage settling of the successive stage prior to engine ignition.

A delayed dual plane separation is employed between the S-IC and S-II stages, while a single plane separation is adequate between the S-II and S-IVB stages.

Umbilicals

In the design and placement of vehicle plates, consideration was given to such things as size, locations, methods of attachment, release, and retraction.

The number of umbilicals is minimized by the combining of electrical connectors and pneumatic and propellant couplings into common umbilical carriers. Location of the umbilicals depended upon the location of the vehicle plates, which were limited somewhat by the propellant tanking, plumbing, and wiring runs inside the vehicle structure. Umbilical disconnect and retraction systems are redundant for reasons of reliability and safety.

Electrical Systems

An electrical load analysis of the launch vehicle provided the incorporation of design features which will enable the flight crew to detect and react effectively to abnormal circumstances. This permits the flight crew to abort safely if the condition is dangerous or to continue the normal mission in an alternate mode if crew safety is not involved but equipment is not operating properly.

Failure Effects and Criticality Analyses

The modes of failure for every critical component of each system are identified. The effect of each failure mode on the operation of the system is analyzed, and those parts contributing most to unreliability are identified. These analyses have resulted in the identification of mission compromising, single-point failures, and have aided in the determination of redundancy requirements and/or design changes.

Design Reviews

A systematic design review of every part, component, subsystem, and system has been performed using comprehensive check lists, failure effects analysis, criticality ratings, and reliability predictions. These techniques have enabled the designer to review the design approach for problems not uncovered in previous analyses. In the R & QA area, the preliminary design review (PDR) and critical design review (CDR) required by the Apollo Program Directive No. 6 represents specialized application of this discipline.

VEHICLE DEVELOPMENT FLOW

Principal milestones in the hardware and mission phases of the Apollo program are shown in figure 1-1.

Certification and Review Schedules

Certificates of Flight Worthiness (COFW) function as a certification and review instrument. A COFW is generated for each major piece of flight hardware. The certificate originates at the manufacturing facility, and is shipped with the hardware wherever it goes to provide a time phased historical record of the item's test results, modifications, failures, and repairs.

The MSFC flight readiness review (MSFC-FRR), the countdown demonstration test (CDDT) and the manned spaceflight-flight readiness review (MSF-FRR) provide assessments of launch vehicle, spacecraft and launch facility readiness. During the final reviews, the decision is made as to when deployment of the world wide mission support forces should begin.

TRANSPORTATION

The Saturn stage transportation system provides reliable and economical transportation for stages and special payloads between manufacturing areas, test areas and KSC. The various modes of transportation encompass land, water, and air routes.

Each stage in the Saturn V system requires a specially designed transporter for accomplishing short distance land moves at manufacturing, test, and launch facilities. These transporters have been designed to be compatible with manufacturing areas, dock facility roll-on/roll-off requirements, and to satisfy stage protection requirements.

Long distance water transportation for the Saturn V stages is by converted Navy barges and landing ship dock type ocean vessels. Tie-down systems provide restraint during transit. Ocean vessels are capable of ballasting to mate with barges and dock facilities for roll-on/roll-off loading. Docks are located at MSFC, KSC, Michoud, MTF, and Seal Beach, California (near Los Angeles).

Air transportation is effected by use of a modified Boeing B-377 (Super Guppy) aircraft. This system provides quick reaction time for suitable cargo requiring transcontinental shipments. For ease in loading and unloading the aircraft, compatible ground support lift trailers are utilized.

A Saturn transportation summary is presented in figure 1-2.

LAUNCH VEHICLE DESCRIPTION

GENERAL ARRANGEMENT

The Saturn V/Apollo general configuration is illustrated in figure 1-3. Also included are tables of engine data, gross vehicle dimensions and weights, ullage and retrorocket data, and stage contractors.

INTERSTAGE DATA FLOW

In order for the Saturn V launch vehicle and Apollo spacecraft to accomplish their objectives, a continuous flow of data is necessary throughout the vehicle. Data flow is in both directions: from spacecraft to stages, and from stages to the spacecraft. The IU serves as a central data processor, and nearly all data flows through the IU.

Specific data has been categorized and tabulated to reflect, in figure 1-4, the type of data generated, its source and its flow. Each stage interface also includes a confidence loop, wired in series through interstage electrical connectors, which assures the Launch Vehicle Digital Computer (LVDC) in the IU that these connectors are mated satisfactorily.

RANGE SAFETY AND INSTRUMENTATION

GENERAL

In view of the hazards inherent in missile/space vehicle programs, certain stringent safety requirements have been established for the Air Force Eastern Test Range (AFETR). Figure 1-5 illustrates the launch azimuth limits and destruct azimuth limits for the Atlantic Missile Range (AMR).

Prime responsibility and authority for overall range safety is vested in the Commander, AFETR, Patrick AFB, Florida. However, under a joint agreement between DOD and NASA, ground safety within the confines of the Kennedy Space Center will be managed by NASA.

To minimize the inherent hazards of the Saturn/Apollo program, a number of safety plans have been developed and implemented in accordance with AFETR regulations.

These plans cover all phases of the Saturn/Apollo program from design, through launch of the vehicle, into orbit.

To enhance the development and implementation of the range safety program, two general safety categories have been established: ground safety and flight safety.

basic data (voltage, frequency, and power requirements) for design of the electrical system.

Such factors as reliability, weight limitations, and weight distributions dictated the requirements to minimize electrical wiring, yet distribute the electrical loads and power sources throughout the launch vehicle. Each stage of the vehicle has its own independent electrical system. No electrical power is transferred between stages; only control signals are routed between stages.

Primary flight power is supplied by wet cell batteries in each stage. The sizes, types, and characteristics are discussed in subsequent sections of this manual. Where alternating current, or direct current with a higher voltage than the batteries is required, inverters and/or converters convert the battery power to the voltages and frequencies needed.

All stages of the launch vehicle are electrically bonded together to provide a unipotential structure, and to minimize current transfer problems in the common side of the power systems.

MANUFACTURE AND LAUNCH CONCEPTS

The development of the vehicle concept required concurrent efforts in the areas of design, manufacture, transportation, assembly, checkout, and launch.

The size and complexity of the vehicle resulted in the decision to have detail design and manufacture of each of the three stages, the Instrument Unit (IU), and the engines accomplished by separate contractors under the direction of MSFC.

This design/manufacturing approach required the development of production plans and controls, and transportation and handling systems capable of handling the massive sections.

The assembly, checkout, and launch of the vehicle required the development of an extensive industrial complex at KSC. Some of the basic ground rules which resulted in the KSC complex described in Section VIII are:

1. The vehicle will be assembled and checked out in a protected environment before being moved to the launch site.

2. A final checkout will be performed at the launch site prior to launch.

3. Once the assembly is complete, the vehicle will be transported in the erect position without disconnecting the umbilicals.

4. Automatic checkout equipment will be required.

5. The control center and checkout equipment will be located away from the launch area.

LAUNCH REQUIREMENTS

Some of the launch requirements which have developed from the application of these ground rules are:

1. Several days prior to the actual launch time, the vehicle is moved to the launch area for prelaunch servicing and checkout. During most of this time, the vehicle systems are sustained by ground support equipment. However, at approximately T-50 seconds, power is transferred to the launch vehicle batteries, and final vehicle systems monitoring is accomplished.

2. While in the launch area, environmental control within the launch vehicle is provided by environmental control systems in the mobile launcher (ML) and on the pad. The IU also utilizes an equipment cooling system, in which heat is removed by circulation of a methanol-water coolant. During preflight, heat is removed from the coolant by a Ground Support Equipment (GSE) cooling system located on the ML. During flight, heat is removed from the coolant by a water sublimator system.

3. While in transit between assembly area and launch area, or while in the launch area for launch preparations, the assembled launch vehicle must withstand the natural environment. The launch vehicle is designed to withstand 99.9% winds during the strongest wind month, while either free standing or under transport, with the damper system attached. In the event of a nearby explosion of a facility or launch vehicle, the Saturn V will also withstand a peak overpressure of 0.4 psi.

4. To more smoothly control engine ignition, thrust buildup and liftoff of the vehicle, restraining arms provide support and holddown at four points around the base of the S-IC stage. A gradual controlled release is accomplished during the first six inches of vertical motion.

RELIABILITY AND QUALITY ASSURANCE

The Apollo Program Office, MA, has the overall responsibility for development and implementation of the Apollo reliability and quality assurance (R & QA) program. NASA Centers are responsible for identifying and establishing R & QA requirements and for implementing an R & QA program to the extent necessary to assure the satisfactory performance of the hardware for which they are responsible. The Apollo R & QA program is defined by the Apollo Program Development Plan, M-D MA 500 and Apollo R & QA Program Plan, NHB 5300-1A.

Crew safety and mission success are the main elements around which the R & QA program is built. The primary criterion governing the design of the Apollo system is that of achieving mission success without unacceptable risk of life or permanent physical disablement of the crew.

It is Apollo program policy to use all currently applicable methods to ensure the reliability and quality of Apollo/Saturn systems. Some of these methods are discussed in subsequent paragraphs.

Analysis of Mission Profiles

The mission profile is analyzed to determine the type and scope of demands made on equipment and flight crew during each phase of the mission. This has resulted in the

Figure 1-1

Figure 1-2

GROUND SAFETY

The ground safety program includes a ground safety plan which calls for the development of safety packages. The major categories covered by these packages are:

1. Vehicle Destruct System. This package includes a system description, circuit descriptions, schematics, ordnance system description, specifications, RF system description, installation, and checkout procedures.

2. Ordnance Devices. This package includes descriptive information on chemical composition and characteristics, mechanical and electrical specifications and drawings, and electrical bridgewire data.

3. Propellants. This package includes descriptive data on chemical composition, quantities of each type, locations in the vehicle, handling procedures, and hazards.

4. High Pressure Systems. This package includes types of gases, vehicle storage locations, pressures, and hazards.

5. Special Precautionary Procedures. This package covers possible unsafe conditions and includes lightning safeguards, use of complex test equipment, and radiological testing.

Also included under ground safety are provisions for launch area surveillance during launch activities. Surveillance methods include helicopters, search radars, and range security personnel. Automatic plotting boards keep the range safety officer (RSO) informed of any intrusion into the launch danger zones by boats or aircraft.

To further assist the RSO in monitoring launch safety, a considerable amount of ground instrumentation is used. A vertical-wire sky screen provides a visual reference used during the initial phase of the launch to monitor vehicle attitude and position. Television systems photographing the launch vehicle from different angles also provide visual reference. Pulsed and CW tracking radars and real time telemetry data provide an electronic sky screen, which displays on automatic plotting boards, and charts the critical flight trajectory parameters.

In the event that the launch vehicle deviates from its planned trajectory, to the degree that it will endanger life or property, the RSO must command destruct by means of the range

1-5

SATURN V LAUNCH VEHICLE

SOLID ULLAGE ROCKET AND RETROROCKET SUMMARY

STAGE	TYPE	QUANTITY	NOMINAL THRUST AND DURATION	PROPELLANT GRAIN WEIGHT
S-IC	RETROROCKET	8	75,800 POUNDS • 0.541 SECONDS	278.0 POUNDS
S-II	ULLAGE	4	23,000 POUNDS † 3.75 SECONDS	336.0 POUNDS
	RETROROCKET	4	34,810 POUNDS ‡ 1.52 SECONDS	268.2 POUNDS
S-IVB	ULLAGE	2	3,390 POUNDS ** 3.87 SECONDS	58.8 POUNDS

ENGINE DATA

STAGE	QTY	ENGINE MODEL	NOMINAL THRUST EACH	NOMINAL THRUST TOTAL	BURN TIME (MINUTES)
S-IC	5	F-1	1,530,000	7,650,000 ††	2.8
S-II	5	J-2	230,000	1,150,000	6.1
S-IVB	1	J-2	207,000	207,000	1ST 2.4 2ND 5.6

STAGE DIMENSIONS / STAGE WEIGHTS

	DIAMETER	LENGTH	DRY	AT LAUNCH
S-IC Base (including fins)	63.0 FEET	138 FEET	288,650 POUNDS	5,022,262 POUNDS
S-IC Mid-stage	33.0 FEET			
S-II Stage	33.0 FEET	81.5 FEET	80,220 POUNDS	1,060,415 POUNDS
S-IVB Stage	21.7 FEET	59.3 FEET	25,050 POUNDS	261,578 POUNDS
Instrument Unit	21.7 FEET	3.0 FEET	4,306 POUNDS	4,306 POUNDS

SATURN V STAGE MANUFACTURERS

STAGE	MANUFACTURER
S-IC	THE BOEING COMPANY
S-II	NORTH AMERICAN-ROCKWELL
S-IVB	McDONNELL - DOUGLAS CORP.
S-IU	INTERNATIONAL BUSINESS MACHINE CORP.

NOTE: THRUST VALUES, WEIGHTS, AND BURN TIMES ARE ALL APPROXIMATIONS.

IU

S-IVB STAGE

363 FEET

S-II STAGE

S-IC STAGE

PRE-LAUNCH LAUNCH VEHICLE GROSS WEIGHT ≈ 6,348,659 POUNDS

• MINIMUM VACUUM THRUST AT 120°F
† AT 170,000 FT. AND 70°F
‡ NOMINAL VACUUM THRUST AT 60°F
** AT 175,000 FT AND 70°F
†† AT SEA LEVEL

Figure 1-3

STAGE ELECTRICAL INTERFACE FLOW

SPACECRAFT TO IU

+28 VDC TO EDS
LV ENGINES CUTOFF TO EDS
ATTITUDE ERROR SIGNAL
Q-BALL PITCH AND YAW
S-IVB ENGINE CUTOFF
AGC COMMAND POWER
S-IVB IGNITION SEQUENCE
   START
AUTO ABORT DEACTIVATE   Ⓜ
INITIATE S-II/S-IVB
   SEPARATION   Ⓜ
SPACECRAFT CONTROL
   DISCRETE   Ⓜ
TRANSLUNAR INJECTION
   INHIBIT   Ⓜ

Ⓜ = MANUALLY INITIATED

IU TO SPACECRAFT

EDS LIFTOFF
EDS AUTO ABORT
+28 VDC FOR EDS
+28 VDC FOR Q BALL
S-IVB ULLAGE THRUST OK
GUIDANCE REFERENCE RELEASE
AGC LIFTOFF
Q BALL TEMPERATURE SENSING
S-II AND S-IVB FUEL TANK
   PRESSURE   Ⓥ
LV ATTITUDE REFERENCE
   FAILURE   Ⓥ
LV RATE EXCESSIVE   Ⓥ
EDS ABORT REQUEST   Ⓥ
S-II START/SEPARATION   Ⓥ
STAGE ENGINES OUT   Ⓥ

Ⓥ = VISUALLY DISPLAYED

S-IVB TO IU

+28 VDC FOR TIMING
SWITCH SELECTOR ADDRESS
   VERIFICATION
ENGINE ACTUATOR POSITIONS
ATTITUDE CONTROL RATE GYROS
   SIGNALS
ATTITUDE CONTROL ACCELEROMETER
   SIGNALS
LOX TANK PRESSURE
FUEL TANK PRESSURE
RSCR & PD EBW FIRING UNIT
ARM AND ENGINE CUTOFF ON
ENGINE THRUST OK
TELEMETRY SIGNALS

S-II TO S-IVB

+28 VDC FOR RETRO-ROCKET
   PRESSURE TRANSDUCER
S-IVB ENGINE START ENABLE

S-II TO IU

ENGINE ACTUATOR POSITIONS
+28VDC FOR TIMING
S-IC STAGE SEPARATED
AFT INTERSTAGE SEPARATED
S-II STAGE SEPARATED
S-II ENGINE OUT
S-II PROPELLANT DEPLETION
SWITCH SELECTOR VERIFY
FUEL TANK PRESSURE
ENGINE THRUST OK
LOX TANK PRESSURE

IU TO STAGES

STAGE ENGINE ACTUATOR COMMANDS
STAGE ENGINE ACTUATOR MEASURING
   VOLTAGES
+28 VDC FOR SWITCHING AND
   TIMING
STAGE SWITCH SELECTOR SIGNALS
(VERIFY, COMMAND, ADDRESS,
   READ, RESET, ENABLE)
STAGE EDS COMMAND ENGINES OFF
S-IVB ATTITUDE CONTROL SYSTEM
   COMMANDS
TELEMETRY CLOCK AND SYNC.

S-IC TO IU

ATTITUDE CONTROL ACCELEROMETER
   SIGNALS
ATTITUDE CONTROL RATE GYRO
   SIGNALS
+28 VDC FOR TIMING
ENGINES OUT
OUTBOARD ENGINE CUTOFF
S-II ENGINES START ENABLE
SWITCH SELECTOR ADDRESS
   VERIFY
S-IC THRUST OK

Figure 1-4

Figure 1-5

safety command system. The range safety system is active until the vehicle has achieved earth orbit, after which the destruct system is deactivated (safed) by command from the ground.

FLIGHT SAFETY

Flight safety planning began during the conceptual phases of the program. One of the requirements of the range safety program is that, during these early phases, basic flight plans be outlined and discussed and, prior to launch, a final flight plan be submitted and approved. As the program develops, the flight planning is modified to meet mission requirements. The flight plan is finalized as soon as mission requirements become firm.

In addition to the normal trajectory data given in the flight plan, other trajectory data is required by the AFETR. This data defines the limits of normality, maximum turning capability of the vehicle velocity vector, instant impact point data, drag data for expended stages and for pieces resulting from destruct action, and location and dispersion characteristics of impacting stages.

In the event the RSO is required to command destruct the launch vehicle, he will do so by manually initiating two separate command messages. These messages are transmitted to the launch vehicle over a UHF radio link. The first message shuts off propellant flow and results in all engines off. As the loss of thrust is monitored by the EDS, the ABORT light is

turned on in the Command Module (CM). Upon monitoring a second abort cue, the flight crew will initiate the abort sequence. The second command from the RSO is for propellant dispersion, and explosively opens all propellant tanks.

Each powered stage of the launch vehicle is equipped with dual redundant command destruct antennae, receivers, decoders, and ordnance to ensure positive reaction to the destruct commands. To augment flight crew safety, the EDS monitors critical flight parameters. Section III provides a more detailed discussion of the EDS.

PERCEPTIBLE PRELAUNCH EVENTS

Prelaunch events which occur subsequent to astronaut loading (approximately T-3 hours), and which may be felt or heard by the flight crew inside the spacecraft, are identified in figure 1-6. Other events, not shown, combine to create a relatively low and constant background. This background noise includes the sounds of environmental control, propellant replenishment, control pressure gas supplies, propellant boiloff and low pressure, low volume purges.

Significant noises and vibrations may be caused by the starting or stopping of an operation or they may result from turbulent flow of gases or liquids. Figure 1-6 illustrates those events most likely to be heard or felt above the background noise or vibration. At approximately T-8.9 seconds all sounds are hidden by the ignition of the S-IC stage engines.

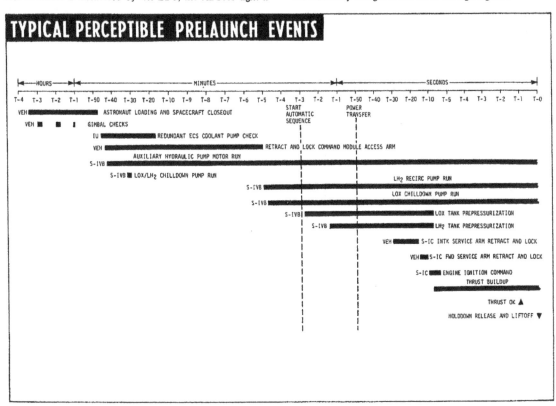

Figure 1-6

This Space Intentionally Left Blank

PERFORMANCE

TABLE OF CONTENTS

INTRODUCTION

Saturn V launch vehicle performance characteristics, under the constraints established by environment and mission requirements, are described in this section. Mission profile, variables, requirements and constraints are described in Section X. The major portion of the performance data presented herein was derived from the AS-507 flight simulations. This data is representative of corresponding data for the AS-508 and AS-509 missions.

FLIGHT SEQUENCE

The space vehicle will be launched from Launch Complex 39 (LC-39) at the Kennedy Space Center. A typical sequence of critical launch events for a nominal mission is contained in figure 2-1. Launch vehicle flight sequence phases are described in the following paragraphs.

LAUNCH AND BOOST TO EARTH PARKING ORBIT

The vehicle rises nearly vertically from the pad, for approximately 450 feet, to clear the tower. During this period, a yaw maneuver is executed to provide tower clearance in the event of adverse wind conditions, deviation from nominal flight and/or engine failure. (See figure 2-1 for start and stop times for this and other maneuvers and events). After clearing the tower, a tilt and roll maneuver is initiated to achieve the flight attitude and proper orientation for the selected flight azimuth. Launch azimuth is 90 degrees; flight azimuth may vary between 72 and 108 degrees. The actual flight azimuth within this range is computed just prior to launch from a launch-day-dependent polynomial of launch azimuth as a function of launch time. From the end of the tilt maneuver to tilt-arrest, the pitch program provides a near-zero-lift trajectory that satisfies vehicle performance, heating and loads requirements. Tilt-arrest freezes the pitch attitude to dampen out pitch rates prior to S-IC/S-II separation. The pitch attitude remains constant until initiation of the Iterative Guidance Mode (IGM) which occurs during the S-II stage flight. Figure 2-2 shows the pitch attitude profile from first motion to earth parking orbit (EPO). Mach 1 is achieved approximately 1 minute 7 seconds after first motion. Maximum dynamic pressure is encountered at approximately 1 minute 24 seconds after first motion. S-IC center engine cutoff is commanded at 2 minutes 15 seconds after first motion, to limit the vehicle acceleration to a nominal 3.98 g. The S-IC outboard engines are cut off by propellant-depletion sensors.

A time interval of 4.4 seconds elapses between S-IC cutoff and the time the J-2 engines of the S-II stage reach the 90%

operating thrust level. During this period, ullage rockets are fired to seat the S-II propellant, the S-IC/S-II separation occurs and the retrorockets back the S-IC stage away from the flight vehicle. Threshold for engine status light OFF is 65% thrust. The S-II aft interstage is jettisoned 30.7 seconds after S-IC cutoff, and the LET is jettisoned by crew action approximately 6 seconds later, after assurance that S-II ignition and thrust buildup have occurred. IGM is initiated about 39 seconds after S-II ignition.

An early S-II center engine cutoff is programmed for Time Base 3 + 299 seconds . This early cutoff is designed to preclude the vehicle longitudinal oscillations (Pogo effect) which have occurred late in the S-II burn in previous missions. The outboard engines are cut off simultaneously when any two of five cutoff sensors in either tank are activated.

An interval of 6.5 seconds elapses between S-II cutoff and the time the S-IVB J-2 engine attains 90% operating thrust level (mainstage). During this coast period, the S-IVB ullage rockets are fired to seat the stage propellant, the S-II/S-IVB separation occurs, and retrorockets back the S-II stage away from the flight vehicle. Threshold for engine status light OFF is 65% thrust. The S-IVB first burn inserts the vehicle into a 100-nautical mile altitude circular parking orbit.

CIRCULAR EARTH PARKING ORBIT

At first S-IVB engine cutoff, the 70-pound thrust auxiliary propulsion system (APS) engines are started and operated for approximately 87 seconds. The LH_2 propulsive vents open approximately 49 seconds after insertion and provide a continuous, low-level thrust to keep the S-IVB propellant seated against the aft bulkheads.

The vehicle coasts in earth parking orbit for up to three revolutions while launch vehicle and spacecraft subsystems checkout is performed. The time to initiate restart preparations for the S-IVB second burn is established by restart geometry criteria.

TRANSLUNAR INJECTION BOOST

The translunar injection boost is part of an ordered flight sequence that begins at initiation of the preignition sequence. The flight computer signals the beginning of the preignition sequence when it determines that the vehicle position satisfies a predesignated geometrical relationship with the target vector. At this time the computer resets to Time Base 6. If a translunar injection inhibit signal from the CM is not sensed, the computer issues the signals that lead to S-IVB reignition. These signals include start helium heater (O_2H_2 burner), close LH_2 tank continuous vent valves, ignite APS ullage engines, restart S-IVB J-2 engine and cutoff ullage engines.

During the preignition sequence, thrust from the continuous LH_2 vent keeps the propellants seated until O_2H_2 burner ignition. The vent is then closed to enable the burner to pressurize both the lox and LH_2 propellant tanks.

TYPICAL CRITICAL EVENT SEQUENCE, FIRST OPPORTUNITY TLI
(EVENT TIMES ARE BASED ON AS-507 SIMULATIONS FOR MISSION H-1
NOVEMBER 14, 1969 WINDOW, 72.029° FLIGHT AZIMUTH)

TIME FROM FIRST MOTION (HR:MIN:SEC)	TIME FROM REFERENCE (HR:MIN:SEC)	EVENT	TIME FROM FIRST MOTION (HR:MIN:SEC)	TIME FROM REFERENCE (HR:MIN:SEC)	EVENT
-0:00:17.3	T_1-0:00:17.7	Guidance Reference Release	0:12:28.8	T_5+0:00:59.0	LH$_2$ Continuous Vent Open
0:00:00.0	T_1-0:00:00.4	First Motion	0:12:56.9	T_5+0:01:27.1	S-IVB APS Ullage Cutoff
0:00:00.4	T_1+0:00:00.0	Liftoff	0:13:18.3	T_5+0:01:48.5	Begin Orbital Navigation Calculations
0:00:01.4	T_1+0:00:01.0	Begin Tower Clearance Yaw Maneuver			
0:00:09.4	T_1+0:00:09.0	End Yaw Maneuver	2:37:41.8	T_6+0:00:00.0	Begin S-IVB Restart Preparations
0:00:12.3	T_1+0:00:11.9	Pitch and Roll Initiation	2:38:23.8	T_6+0:00:42.0	O$_2$H$_2$ Burner (Helium Heater) On
0:01:06.9	T_1+0:01:06.5	Mach 1	2:38:24.0	T_6+0:00:42.2	LH$_2$ Continuous Vent Closed
0:01:23.5	T_1+0:01:23.1	Maximum Dynamic Pressure	2:45:58.1	T_6+0:08:16.3	S-IVB APS Ullage Ignition
0:02:15.0	T_2+0:00:00.0	S-IC Center Engine Cutoff	2:45:58.6	T_6+0:08:16.8	Helium Heater Off
0:02:38.3	T_2+0:00:23.3	Begin Tilt Arrest	2:47:11.8	T_6+0:09:30.0	S-IVB Engine Restart Sequence
0:02:42.2	T_3+0:00:00.0	S-IC Outboard Engine Cutoff	2:47:14.9	T_6+0:09:33.1	S-IVB APS Ullage Cutoff
0:02:42.7	T_3+0:00:00.5	S-II Ullage Rocket Ignition	2:47:19.8	T_6+0:09:38.0	S-IVB Reignition (Start Tank Discharge Valve Opens)
0:02:42.9	T_3+0:00:00.7	Signal to Separation Devices and S-IC Retrorockets	2:47:22.3	T_6+0:09:40.5	S-IVB Engine at 90% Thrust
0:02:43.0	T_3+0:00:00.8	S-IC/S-II First Plane Separation Complete	2:49:02.3	T_6+0:11:20.5	MR Shift (First Opportunity Only)
0:02:43.6	T_3+0:00:01.4	S-II Engine Start Sequence Initiated	2:53:04.8	T_7-0:00:00.2	S-IVB Engine Cutoff, Second Burn
0:02:44.6	T_3+0:00:02.4	S-II Ignition (Start Tank Discharge Valve Opens)	2:53:05.0	T_7+0:00:00.0	Set Time Base 7
0:02:46.6	T_3+0:00:04.4	S-II Engines at 90% Thrust	2:53:05.5	T_7+0:00:00.5	LH$_2$ Continuous Vent Open
0:02:47.2	T_3+0:00:05.0	S-II Ullage Thrust Cutoff	2:53:05.7	T_7+0:00:00.7	Lox Nonpropulsive Vent Open
0:03:12.9	T_3+0:00:30.7	S-II Aft Interstage Drop (Second Plane Separation)	2:53:05.8	T_7+0:00:00.8	LH$_2$ Nonpropulsive Vent Open
0:03:18.4	T_3+0:00:36.2	LET Jettison (Crew Action)	2:53:08.6	T_7+0:00:03.6	Flight Control Coast Mode On
0:03:23.0	T_3+0:00:40.8	Initiate IGM	2:53:10.0	T_7+0:00:05.0	Enable SC Control of LV
0:04:22.2	T_2+0:01:40.0	S-II Lox Tank Pressurization Flowrate Step	2:53:14.8	T_7+0:00:09.8	Translunar Injection
0:07:41.2	T_3+0:04:58.0	S-II Center Engine Cutoff	2:53:25.0	T_7+0:00:20.0	Initiate Maneuver to and Maintain Local Horizontal Alignment (CSM Forward, Heads Down)
0:07:42.2	T_3+0:05:00.0	S-II Fuel Tank Pressurization Flowrate Step			
0:09:10.7	T_4+0:00:00.0	S-II Outboard Engine Cutoff	2:55:35.7	T_7+0:02:30.7	Lox Nonpropulsive Vent Closed
0:09:11.5	T_4+0:00:00.8	S-IVB Ullage Ignition	3:08:04.8	T_7+0:14:59.0	LH$_2$ Nonpropulsive Vent Closed
0:09:11.6	T_4+0:00:00.9	Signal to Separation Devices and S-II Retrorockets	3:08:04.8	T_7+0:14:59.8	LH$_2$ Continuous Vent Closed
0:00:11.7	T_4+0:00:01.0	S-II/S-IVB Separation	3:08:05.0	T_7+0:15:00.0	Initiate Maneuver to and Maintain TD&E Attitude
0:09:11.7	T_4+0:00:01.0	S-IVB Engine Start Sequence, First Burn	3:18:05.0	T_7+0:25:00.0	CSM Separation (Variable)
0:09:14.7	T_4+0:00:04.0	S-IVB Ignition (Start Tank Discharge Valve Opens)	3:28:05.0	T_7+0:35:00.0	CSM/LM Docking (Variable)
0:09:17.2	T_4+0:00:06.5	S-IVB Engine at 90% Thrust	3:53:05.4	T_7+1:00:00.4	LH$_2$ Nonpropulsive Vent Open
0:09:19.3	T_4+0:00:08.6	S-IVB Ullage Thrust End	4:08:04.0	T_7+1:14:59.0	LH$_2$ Nonpropulsive Vent Closed
0:09:23.5	T_4+0:00:12.8	S-IVB Ullage Case Jettison	4:18:05.0	T_7+1:25:00.0	SC/LV Final Separation (Variable)
0:11:23.1	T_4+0:02:12.4	Begin Chi Freeze	4:21:05.0	T_7+1:28:00.0	Initiate Maneuver to and Maintain S-IVB Evasive Attitude (Variable)
0:11:29.6	T_5-0:00:00.2	S-IVB Cutoff, First Burn			
0:11:29.8	T_5+0:00:00.0	Set Time Base 5	4:29:05.0	T_8+0:00:00.0	Set Time Base 8
0:11:30.1	T_5+0:00:00.3	S-IVB APS Ullage Ignition	4:29:06.2	T_8+0:00:01.2	S-IVB APS Ullage Ignition
0:11:39.6	T_5+0:00:09.8	Parking Orbit Insertion	4:30:26.2	T_8+0:01:21.2	S-IVB APS Ullage Cutoff
0:11:49.8	T_5+0:00:20.0	Initiate Maneuver to and Maintain Local Horizontal Alignment (CSM Forward, Heads Down)	4:38:45.0	T_8+0:09:40.0	Initiate Maneuver to and Maintain Slingshot Attitude
			4:38:45.2	T_8+0:09:40.2	LH$_2$ Continuous Vent Open
0:11:50.3	T_5+0:00:20.5	Begin Orbital Guidance	4:50:25.0	T_8+0:21:20.0	Start Lox Dump
			4:51:23.0	T_8+0:22:18.0	End Lox Dump
			4:52:25.2	T_8+0:23:20.2	Lox Nonpropulsive Vent Open
			4:52:30.0	T_8+0:23:25.0	LH$_2$ Nonpropulsive Vent Open
			5:25:45.0	T_8+0:56:40.0	S-IVB APS Ullage Ignition
			5:30:45.0	T_8+1:01:40.0	S-IVB APS Ullage Cutoff
			5:39:05.0	T_8+1:10:00.0	Initiate Maneuver to and Maintain Communications Attitude

Figure 2-1

Two opportunities for translunar injection are provided. For first injection opportunity, S-IVB reignition occurs after approximately 1.5 revolutions in parking orbit (Pacific window). The second opportunity occurs after 2.5 revolutions in parking orbit.

COAST IN TRANSLUNAR ORBIT

Following the S-IVB cutoff, the translunar orbit is perturbed by a propulsive hydrogen vent that lasts for 15 minutes. During this period, lox and LH_2 nonpropulsive vents are also operative. After the vents are closed, the vehicle maneuvers to the commanded transposition, docking, and extraction (TD&E) attitude and this attitude is maintained inertially throughout TD&E operations. (See figure 10-3, Section X).

S-IVB EVASIVE MANEUVER

Following the separation of the spacecraft (combined CSM/LM) from the launch vehicle (S-IVB/IU), a maneuver inhibit is released by ground command and the S-IVB/IU maneuvers to the evasive maneuver attitude. The evasive maneuver attitude is the same as the TD&E inertial attitude except that the yaw attitude angle changes in sign but not in magnitude, for example, the yaw angle changes from -40 to +40 degrees; pitch and roll angles remain the same. (See figure 10-3, Section X). After the S-IVB/IU has achieved and is holding the evasive maneuver attitude, and this fact is confirmed by the crew, Time Base 8 is initiated by ground command. An 80-second burn of the S-IVB APS ullage engines provides a velocity increment of 10 feet per second and propels the S-IVB/IU away from the spacecraft.

SLINGSHOT MODE

After spacecraft ejection from the launch vehicle and the S-IVB evasive maneuver are completed, the slingshot mode is initiated (T_8 +580 seconds). The slingshot procedure is designed to minimize the probability of spacecraft recontact with the launch vehicle, S-IVB earth impact, and S-IVB lunar impact, in that order of priority. The slingshot mode is achieved by maneuvering the S-IVB stage to the slingshot attitude and performing a retrograde dump of the residual lox to decrease the velocity of the stage. Propulsive hydrogen venting and an APS burn also contribute to the retrograde velocity increment. The decrease in velocity perturbs the trajectory so that the vehicle coasts past the trailing side of the moon and thus, is co-rotational with the moon. The moon's gravitational field increases the launch vehicle's velocity sufficiently to place it into solar orbit (see figure 10-1, Section X).

The S-IVB mass during the slingshot propulsive operations varies from launch day to launch day and across each launch window. This variation in mass causes the slingshot velocity differential to be variable. The slingshot attitude, which is maintained relative to the local horizontal coordinate system throughout slingshot operations, also varies with the launch date. For the November 14, 1969 launch day simulations, the slingshot velocity increment is approximately 102 feet per second and the slingshot attitude angles in pitch, yaw and roll are 191, 0, and 180 degrees, respectively.

Following the propellant dump, the S-IVB Stage is "safed" by venting the remaining propellant and high pressure gas bottles.

Upon completion of slingshot operations, the S-IVB/IU maneuvers to and maintains an attitude suitable for communication with ground stations.

FLIGHT PERFORMANCE

The typical flight performance data presented herein are based on launch vehicle operational trajectory studies. These studies were based on the requirements and constraints imposed by the mission.

FLIGHT PERFORMANCE PARAMETERS

Flight performance parameters for the mission are presented graphically in figures 2-2 through 2-21. These parameters are shown for nominal cases for the earth parking orbit insertion and TLI phases. Parameters shown include pitch angle, vehicle weight, axial force, aerodynamic pressure, longitudinal acceleration, inertial velocity, altitude, range, angle of attack, inertial path angle and inertial azimuth.

FLIGHT PERFORMANCE AND FLIGHT GEOMETRY PROPELLANT RESERVES

Required propellant reserves are comprised of two components: Flight Performance Reserves (FPR) and Flight Geometry Reserves (FGR). The FPR is defined as the root-sum-square (RSS) combination of negative launch vehicle weight dispersions at TLI due to 3-sigma launch vehicle subsystems and environmental perturbations. The FGR is defined as the reserve propellant required to guarantee the launch vehicle capability to establish a lunar flyby trajectory at any earth-moon geometry. The total reserves required to provide 99.865% assurance that the launch vehicle will complete its primary mission objective is the algebraic sum of the FPR and the FGR.

Predicted propellant reserves are determined from flight simulations and operational trajectory data. These predicted reserves must equal or exceed the required reserves established for a mission.

PROPULSION PERFORMANCE

The typical propulsion performance data presented herein are based on flight simulations, stage and engine configuration, and static test firing data.

PROPELLANT LOADING

A propellant weight summary for each stage is tabulated in figures 2-22 through 2-24. The tables break down propellant use into such categories as usable, unusable, trapped, buildup and holddown, mainstage, thrust decay, and fuel bias.

ENGINE PERFORMANCE

Stage thrust versus time history for the three launch vehicle propulsive stages are graphically presented in figures 2-25 through 2-27.

The predicted thrust profile for the S-IC stage (figure 2-25) shows the thrust increase from approximately 7,693,000 pounds at one second after first motion to approximately 9,042,000 pounds at center engine cutoff, where the vehicle has attained an altitude of approximately 144,000 feet. At center engine cutoff vehicle thrust drops to approximately 7,120,000 pounds.

The S-II stage predicted thrust profile (figure 2-26) is slightly perturbed by the aft interstage drop and launch escape tower jettison. A thrust level of approximately 1,163,000 pounds is

Figure 2-2

Figure 2-3

Figure 2-4

Figure 2-5

Figure 2-6

Figure 2-7

Figure 2-8

Figure 2-9

Figure 2-10

Figure 2-11

Figure 2-12

Figure 2-13

Figure 2-14

Figure 2-15

TYPICAL ANGLE OF ATTACK DURING BOOST TO EPO

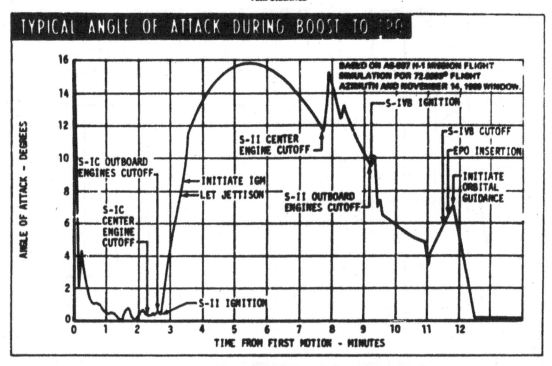

Figure 2-16

TYPICAL ANGLE OF ATTACK DURING TLI BOOST

Figure 2-17

Figure 2-18

Figure 2-19

Figure 2-20

Figure 2-21

attained with a 5.5 mixture ratio (MR). At center engine cutoff the thrust level drops to 925,000 pounds. At the MR shift to 4.3, which occurs after center engine cutoff, the thrust level drops to 690,000 pounds.

The S-IVB stage predicted thrust profiles for first and second burns are shown in figure 2-27. The thrust level for first burn is approximately 207,000 pounds attained with a 5.0 MR. For the first opportunity TLI, the S-IVB burn is started at a 4.5 MR and shifted to 5.0 approximately 100 seconds after

mainstage. For second opportunity TLI, the MR shift from 4.5 to 5.0 occurs 2.5 seconds after mainstage. The longer delay in MR shift for the first opportunity mission is designed to consume the fuel that would have boiled off during a third revolution in earth orbit. The thrust levels for the 5.0 and 4.5 mixture ratio are, respectively, approximately 207,000 and 178,000 pounds.

Thrust level during earth parking orbit is shown in figure 2-28.

S-IC STAGE PROPELLANT WEIGHT SUMMARY		
BASED ON AS-507 PREDICTED DATA	LOX (POUNDS)	RP-1 (POUNDS)
CONSUMED PROPELLANT	3,275,945	1,396,158
BUILDUP AND HOLDDOWN	67,111	18,339
MAINSTAGE	3,195,006	1,368,311
THRUST DECAY	5,441	3,393
TAILOFF	1,635	415
FUEL BIAS	NONE	5,700
PRESSURIZATION	6,752	NONE
RESIDUAL PROPELLANT	32,915	23,303
TANKS	2,160	9,971
SUCTION LINES	28,265	6,449
INTERCONNECT LINES	330	NONE
ENGINES	2,160	6,585
ENGINE CONTROL SYSTEMS	NONE	298
TOTAL	3,308,860	1,419,461

Figure 2-22

S-II STAGE PROPELLANT WEIGHT SUMMARY		
BASED ON AS-507 PREDICTED DATA	LOX (POUNDS)	LH$_2$ (POUNDS)
USABLE PROPELLANT	813,605	155,440
MAINSTAGE	807,920	151,919
BIAS	NONE	1,725
THRUST BUILDUP	1,002	523
THRUST DECAY	225	112
PRESSURIZATION GAS	4,458	1,161
UNUSABLE PROPELLANT	3,533	2,868
TRAPPED:	3,185	2,661
ENGINE AND LINES	1,625	231
INITIAL ULLAGE MASS	338	75
TANK AND SUMP	1,222	2,355
VENTED GAS	348	207
TOTAL	817,138	158,308

Figure 2-23

S-IVB STAGE PROPELLANT WEIGHT SUMMARY				
BASED ON AS-507 PREDICTED DATA	FIRST OPPORTUNITY TLI		SECOND OPPORTUNITY TLI	
	LOX (POUNDS)	LH$_2$ (POUNDS)	LOX (POUNDS)	LH$_2$ (POUNDS)
TOTAL USABLE PROPELLANT	188,754	39,618	188,752	38,766
USABLE	188,754	39,202	188,752	38,350
(INCLUDES MAINSTAGE FLIGHT PERFORMANCE AND FLIGHT GEOMETRY RESERVES)				
RESIDUAL FUEL BIAS	NONE	416	NONE	416
UNUSABLE PROPELLANT	1,420	3,882	1,422	4,734
ORBITAL BOILOFF	142	2,709	151	3,563
*FUEL LEAD	NONE	37	NONE	37
MAINSTAGE BOILOFF	18	50	12	48
ENGINE TRAPPED	108	10	108	10
LINES TRAPPED	259	38	259	38
TANK UNAVAILABLE	40	685	40	685
*BUILDUP TRANSIENTS	702	264	701	264
*DECAY TRANSIENTS	139	61	139	61
REPRESSURIZATION (O$_2$H$_2$ BURNER)	12	28	12	28
* FOR FIRST AND SECOND BURNS				
TOTAL	190,174	43,500	190,174	43,500

Figure 2-24

Figure 2-25

Figure 2-26

Figure 2-27

Figure 2-28

FLIGHT LOADS

Flight loads are dependent on the flight trajectory, associated flight parameters, and wind conditions. These factors are discussed in the following paragraphs. These loads based on AS-507 predictions, are representative of corresponding loads for the AS-508 and AS-509 missions.

HIGH DYNAMIC PRESSURE/WIND LOADS

The launch vehicle bending moments through the high q region are dependent on the shape of the wind profile and the orientation of the wind vector with respect to the trajectory plane. The variations of maximum launch vehicle bending moment with altitude of peak wind speed and wind azimuth have been evaluated based on the September through December 95 percentile directional winds. These variations are shown in figures 2-29 and 2-30. The critical wind azimuth for the directional synthetic profile was 300 degrees measured clockwise from true North. A gust thickness of 640 feet was superimposed on the synthetic wind profile that peaked at an altitude of 32,800 feet with a wind velocity of 174 feet per second.

The maximum bending moment diagram due to flight through the above winds is shown in figure 2-31. A maximum bending moment of 149 million inch-pounds occurs at about station 1100 at 72.4 seconds after first motion.

CENTER ENGINE CUTOFF LOADS

S-IC center engine cutoff (CECO) is programmed for 135 seconds after first motion. Figure 2-32 shows the axial load at CECO. The nominal longitudinal load factor at CECO is 3.75 g's.

OUTBOARD ENGINE CUTOFF LOADS

S-IC outboard engine cutoff (OBECO) occurs at approximately 162 seconds after first motion. Axial load at OBECO is shown in figure 2-33. The nominal longitudinal load factor at OBECO is 3.91 g's.

ENGINE-OUT CONDITIONS

Engine-out conditions, if they should occur, will affect the vehicle loads. The time at which the malfunction occurs, which engine malfunctions, peak wind speed and azimuth orientation of the wind, are all independent variables which combine to produce load conditions. Each combination of engine-out time, peak wind velocity, wind azimuth, and altitude at which the maximum wind shear occurs, produces a unique trajectory. Vehicle responses such as dynamic pressure, altitude, Mach number, angle-of-attack, engine gimbal angles, yaw and attitude angle time histories vary with the prime conditions. Structure test programs indicate a positive structural margin exists for this malfunction flight condition.

Studies indicate that the immediate structural dynamic transients at engine-out will not cause structural failure. However, certain combinations of engine failure, wind direction and wind magnitude may result in a divergent control condition which could cause loss of the vehicle.

The "Chi-Freeze" schedule is incorporated into the vehicle

TYPICAL ENVELOPE OF MAXIMUM BENDING MOMENT-WIND ALTITUDE

FOR NOMINAL FLIGHT AT STATION 1156

BASED ON AS-507 PREDICTED DATA

ALTITUDE OF PEAK WIND SPEED - THOUSAND FEET

BENDING MOMENT - MILLION INCH - POUNDS

Figure 2-29

Figure 2-30

Figure 2-31

Figure 2-32

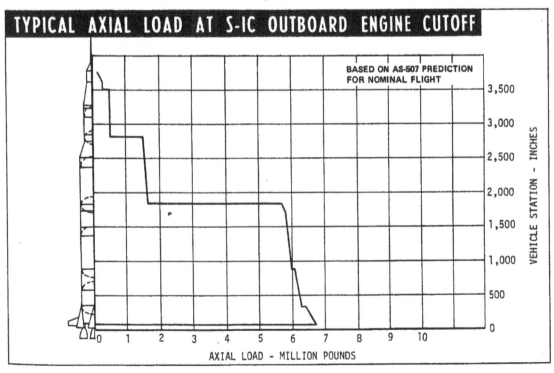

Figure 2-33

guidance program as an alternate to reduce the effect of loss in thrust from an S-IC engine. (Freeze initiation and freeze duration are dependent upon the time at which the loss in thrust occurs.) This schedule holds the pitch attitude command constant, thereby providing a higher altitude trajectory. The higher altitude trajectory minimizes the payload losses into orbit. It also improves the vehicle engine-out dynamic response by providing a lower-velocity entry into the maximum aerodynamic region.

A single control engine-out during S-II powered flight does not produce load conditions which are critical.

LIFTOFF CLEARANCE

Launch vehicle liftoff interference points and the predicted

minimum clearances for the AS-507 mission are shown in figure 2-34. The points listed in the figure are the most restrictive interference points. Clearance at these points guarantees clearance at all other potential interference points.

Liftoff clearances are significantly affected by engine misalignment, stacking errors, actuator valve misalignment, rate gyro error and wind loading.

A yaw bias at liftoff is included in the flight program to move the vehicle away from the tower. This yaw bias is ramped to 1.25 degrees at the rate of 1.0 degrees per second starting at 1.39 seconds from first motion and is ramped back to zero degrees 9 seconds later.

SUMMARY OF LIFTOFF CLEARANCES (AS 507 PREDICTION)			
POTENTIAL INTERFERENCE POINTS		AVAILABLE CLEARANCE (INCHES)	PREDICTED MINIMUM CLEARANCE (INCHES) ①
VEHICLE	GROUND EQUIPMENT		
Thrust Structure	Holddown Post	3.0	1.62
Thrust Structure Insulation	Liftoff Switch Arm	②	③
F-1 Engine	Holddown Post	44.5	10.1
Fin Tip	Swing Arms	②	64.6

① Combined 95 percentile wind and 0.34 degree thrust vector misalignment

② Variable

③ Switch arm roller remains on striker plate

Figure 2-34

EMERGENCY DETECTION AND PROCEDURES

TABLE OF CONTENTS

LAUNCH VEHICLE MONITORING AND CONTROL

The spacecraft is equipped with a number of displays and controls which permit monitoring the launch vehicle conditions and controlling the launch vehicle under normal and emergency conditions. Many of these displays and controls are related to the Emergency Detection System (EDS).

The displays implemented for EDS monitoring were selected to present as near as possible those parameters which represent the failures leading to vehicle abort. Whenever possible, the parameter was selected so that it would display total subsystem operation. Manual abort parameters have been implemented with redundant sensing and display to provide highly reliable indications to the crewmen. Automatic abort parameters have been implemented triple redundant, voted two-out-of-three, to preclude single point hardware or sensing failures causing an inadvertent abort.

The types of displays have been designed to provide onboard detection capability for rapid rate malfunctions which may require abort. Pilot abort action must, in all cases, be based on two separate but related abort cues. These cues may be derived from the EDS displays, ground information, physiological cues, or any combination of two valid cues. In the event of a discrepancy between onboard and ground based instrumentation, onboard data will be used. The EDS displays and controls are shown in figure 3-1. As each is discussed it is identified by use of the grid designators listed on the border of the figure.

LAUNCH VEHICLE MONITORING DISPLAYS

FLIGHT DIRECTOR ATTITUDE INDICATOR

There are two Flight Director Attitude Indicators (FDAI's), each of which provides a display of Euler attitude, attitude errors and angular rates. Refer to figure 3-1, Q-45 and J-61, for locations of the FDAI's on the MDC and to figure 3-2 for details of the FDAI's. These displays are active at liftoff and remain active throughout the mission, except that attitude errors are not displayed during S-II and S-IVB flight. The FDAI's are used to monitor normal launch vehicle guidance and control events. The roll and pitch programs are initiated simultaneously at +10 seconds. The roll program is terminated when flight azimuth is reached, and the pitch program continues to tilt-arrest. IGM initiate will occur approximately five seconds after LET jettison during the S-II stage flight.

The FDAI ball displays Euler attitude, while needle type pointers across the face of the ball indicate attitude errors, and triangular pointers around the periphery of the ball display angular rates. Attitude errors and angular rate displays are, clockwise from the top, roll, pitch and yaw, respectively.

Signal inputs to the FDAI's are switch selectable and can come from a number of different sources in the spacecraft. This flexibility and redundancy provides the required attitude and error backup display capability.

Excessive pitch, roll, or yaw indications provide a single cue that an abort is required. Additional abort cues will be provided by the FDAI combining rates, error, or total attitude. Second cues will also be provided by the LV RATE light (R-50), LV GUID light (R-52), physiological sensations and MCC ground reports.

LV ENGINES LIGHTS

Each of the five LV ENGINES lights (S-51, figure 3-1) represents the respective numbered engine on the operating stage. (e.g., light number one represents engine number one on the S-IC, S-II, or S-IVB stage; light number two represents engine number two on the S-IC, or S-II stage; etc.).

These lights are controlled by switching logic in the IU. The switching logic monitors thrust ok pressure switches on each engine of the operating stage and also staging discretes. A light ON indicates its corresponding engine is operating below a nominal thrust level (90% on F-1 engines and 65% on J-2 engines). During staging, all lights are turned OFF momentarily to indicate that physical separation has occured.

LV RATE LIGHT

The LV RATE light (R-50, figure 3-1) is an abort cue indicator. During the time that the overrate portion of the automatic abort system is enabled (liftoff to approximately T + 2 minutes) the LV RATE light is time shared to indicate either an LV overrate condition or a LV platform failure (second cue). After the automatic abort system is deactivated the LV RATE light is used only to indicate that preset LV overrate settings have been exceeded. The light is a single cue for abort. Secondary cues will be provided by the FDAI, physiological cues, or ground communications.

The LV preset overrate settings are:

Pitch and Yaw	4.0 (±0.5) degrees/sec.	Liftoff to S-II ignition
	9.2 (±0.8) degrees/sec.	S-II ignition to S-IVB cutoff
Roll	20.0 (±0.5) degrees/sec.	Liftoff to S-IVB cutoff

NOTE

The LV RATE light may blink ON and OFF during normal LV staging.

See also LAUNCH VEHICLE EDS CONTROLS for a discussion of EDS switch and LV RATES switch.

MAIN DISPLAY CONSOLE

Figure 3-1. (Sheet 1 of 7)

Figure 3-1. (Sheet 2 of 7)

Figure 3-1. (Sheet 3 of 7)

MAIN DISPLAY CONSOLE

Figure 3-1. (Sheet 4 of 7)

Figure 3-1. (Sheet 5 of 7)

Figure 3-1. (Sheet 6 of 7)

EMERGENCY DETECTION

LEGEND			
ABORT LIGHT	N-51	LIMIT CYCLE SWITCH	T-37
ABORT SYSTEM SWITCHES	AA-59	LOGIC POWER SWITCH	AB-13
ACCELEROMETER	O-40	LES MOTOR FIRE PUSHBUTTON	X-49
ALTIMETER	C-50	LV ENGINE LIGHTS	S-51
APEX COVER JETT PUSHBUTTON	W-52	LV GUID LIGHT	R-52
ATTITUDE SET CONTROL PANEL	AD-32	LV RATE LIGHT	R-50
ATT DEADBAND SWITCH	T-38	LV RATES SWITCH	AA-59
ATT SET SWITCH	R-40	LV/SPS IND SWITCHES	AE-40
AUTO RCS SELECT SWITCHES	P-15	LV TANK PRESS GAUGES	X-46
BMAG MODE SWITCHES	Z-32	LVa/SPS Pc INDICATOR	W-42
BMAG PWR SWITCHES	AA-7	MAIN DEPLOY PUSHBUTTON	Y-52
CANARD DEPLOY PUSHBUTTON	Y-49	MAIN RELEASE SWITCH	AG-56
CMC ATT SWITCH	O-37	MANUAL ATTITUDE SWITCHES	T-34
CMC MODE SWITCH	W-39	MASTER ALARM LIGHT	L-40,O-110
CM PRPLNT SWITCHES	AC-51	PRPLNT DUMP SWITCH	AA-56
CM RCS LOGIC SWITCH	AC-50	RATE SWITCH	T-39
CM RCS He DUMP PUSHBUTTON	Z-52	REACTION CONTROL SYSTEM C/B	U-10
CM/SM SEP SWITCHES	Y-61	ROLL STABILITY INDICATOR	L-45
CSM/LM SWITCH	Y-59	ROT CONTR PWR SWITCHES	W-33
CSM/LV SEP PUSHBUTTON	Z-49	SC CONT SWITCH	W-38
CORRIDOR INDICATORS	K-45,M-45	SCS ELECTRONICS POWER SWITCH	AA-10
DIRECT O$_2$ SWITCH	AD-10	SCS TVC SERVO POWER SWITCHES	AA-14
DIRECT ULLAGE PUSHBUTTON	Z-38	SCS TVC SWITCHES	AC-39
DROGUE DEPLOY PUSHBUTTON	X-52	SERVICE PROPULSION SYSTEM C/B	W-10
DSKY PANEL	R-60	SIG COND/DRIVER BIAS POWER SWITCHES	AC-12
EDS POWER SWITCH	Z-17	SPS THRUST LIGHT	K-49
EDS AUTO SWITCH	Y-57	SPS THRUST SWITCH	Z-36
ELS BATTERY C/B	X-9	SPS GIMBAL THUMBWHEEL CONTROLS	AA-46
ELS SWITCHES	AC-48	SPS GIMBAL MOTORS SWITCHES	AC-42
EMS FUNCTION SWITCH	G-45	STABILITY CONTROL SYSTEM C/B	P-5, R-12,S-9
EMS MODE SWITCH	I-44	S-II SEP LIGHT	R-51
ENTRY SWITCHES	AE-38	S-II/S-IVB LV STAGE SWITCH	AD-58
EDS BATTERY C/B	X-5	S-IVB/LM SEP SWITCH	Y-63
EVENT TIMER INDICATOR	P-51	THRESHOLD INDICATOR (.05 G LIGHT)	K-48
EVENT TIMER SWITCHES	AE-50	THRUST ON PUSHBUTTON	AA-38
FCSM SWITCHES	AE-47	TRANS CONTR SWITCH	T-40
FDAI	J-60, Q-45	TVC GIMBAL DRIVE SWITCHES	AE-42
FDAI CONTROL SWITCHES	R-37	TWR JETT SWITCHES	AA-61
FDAI/GPI POWER SWITCH	AA-12	UP TLM SWITCHES	O-66
GDC ALIGN PUSHBUTTON	AC-36	XLUNAR SWITCH	AD-59
GUIDANCE SWITCH	AC-56	2 ENG OUT SWITCH	AA-58
G-V PLOTTER	H-49	Δ V CG SWITCH	AC-46
IMU CAGE SWITCH	AE-36	Δ V THRUST SWITCHES	Z-41
LAUNCH VEHICLE SWITCHES	AD-58	Δ V/EMS SET SWITCH	K-52
LIFTOFF/NO AUTO ABORT LIGHTS	W-50	Δ V/RANGE COUNTER	L-48

NOTE

THE PANEL INDICATORS AND SWITCHES ASSOCIATED WITH THE EDS AND LAUNCH VEHICLE CONTROL ARE LISTED FOR EASE IN LOCATION. REFER TO THE TEXT FOR DISCUSSION OF SWITCH AND INDICATOR FUNCTIONS.

Figure 3-1. (Sheet 7 of 7)

S-II SEP LIGHT

With S-IC/S-II staging, the S-II SEP light (R-51, figure 3-1) will illuminate. The light will go out approximately 30 seconds later when the interstage structure is jettisoned. A severe overheating problem will occur if the structure is not jettisoned at the nominal time. Under the worst conditions, abort limits will be reached within 25 seconds from nominal jettison time. Confirmation from Mission Control of interstage failure to jettison serves as the second abort cue.

During the earth orbit phase of the mission the S-II SEP light is again used. It is turned ON to indicate the beginning of restart preparations at $T_6 + 0.1$ seconds. It is turned OFF at $T_6 + 41.6$ seconds to indicate O_2H_2 burner ignition. It is turned ON again at $T_6 + 8$ minutes 40 seconds and OFF again at $T_6 + 9$ minutes 20 seconds to indicate translunar injection (TLI) commit (engine start -10 seconds).

WARNING

After TLI commit ($T_6 + 9$ minutes 20 seconds) the only way remaining to effect a TLI inhibit is with the T-handle.

LV GUID LIGHT

The LV platform (ST-124M-3) is interrogated every 25 milliseconds for the correct attitude. If an excessive attitude discrepancy is found during three consecutive checks on the fine resolvers, fifteen additional checks will be performed on the coarse resolvers. (The LV can continue the mission on the coarse resolvers alone.) If the coarse resolvers fail to control within 15 degrees per second rate of attitude change in any plane, commands sent to the flight control system to change LV attitude will be inhibited and the flight control system will hold the last acceptable command.

A signal is sent from the LVDA to activate the LV GUID light (R-52, figure 3-1) at the same time the flight control commands are inhibited. It is a single cue for abort. Second cues will be provided by the LV RATE light (only when the automatic abort system is on) and by the FDAI (Q-45), angle of attack (Qa) meter (W-42) and/or ground information.

The LV GUID light ON is a prerequisite to spacecraft takeover of the Saturn during launch vehicle burn phases. See subsequent paragraph on GUIDANCE SWITCH for a further discussion of spacecraft takeover.

LIFTOFF/NO AUTO ABORT LIGHTS

The LIFTOFF and NO AUTO ABORT lights (W-50, figure 3-1) are independent indications contained in one switch/light assembly.

The LIFTOFF light ON indicates that vehicle release has been commanded and that the IU umbilical has ejected. The spacecraft digital event timer is started by the same function. The LIFTOFF light is turned OFF at S-IC CECO.

The NO AUTO ABORT light ON indicates that one or both of the spacecraft sequencers did not enable automatic abort capability at liftoff. Automatic abort capability can be enabled by pressing the switch/light pushbutton. If the light remains ON one or both of the automatic abort circuits failed to energize. The crew must then be prepared to back up the automatic abort manually. The NO AUTO ABORT light is also turned OFF at S-IC CECO.

WARNING

If the NO AUTO ABORT pushbutton is depressed at T-0 and a pad shutdown occurs, a pad abort will result.

ABORT LIGHT

The ABORT light (N-51, figure 3-1) can be illuminated by ground command from the Flight Director, the Mission Control Center (MCC) Booster Systems Engineer, the Flight Dynamics Officer, the Complex 39 Launch Operations Manager (until tower clearance at +10 seconds), or in conjunction with range safety booster engine cutoff. The ABORT light ON constitutes one abort cue. An RF voice abort request constitutes one abort cue.

NOTE

Pilot abort action is required prior to receipt of an ABORT light or a voice command for a large percentage of the time critical launch vehicle malfunctions, particularly at liftoff and staging.

ANGLE OF ATTACK METER

The angle of attack (Qa) meter (W-42, figure 3-1) is time shared with service propulsion system (SPS) chamber pressure. The Qa display is a pitch and yaw vector summed angle-of-attack/dynamic pressure product (Qa). It is expressed in percentage of total pressure for predicted launch vehicle breakup (abort limit equals 100%). It is effective as an abort parameter only during the high Q flight region from +50 seconds to +2 minutes.

Except as stated above, during ascent, the Qa meter provides trend information on launch vehicle flight performance and provides a secondary cue for slow-rate guidance and control malfunctions. Primary cues for guidance and control malfunctions will be provided by the FDAI, physiological cues, and/or MCC callout.

Nominal angle of attack meter indications should not exceed 25% to 50%. Expected values based on actual winds aloft will be provided by MCC prior to launch.

ACCELEROMETER

The accelerometer (O-40, figure 3-1) indicates longitudinal acceleration/deceleration. It provides a secondary cue for certain engine failures and is a gross indication of launch vehicle performance. The accelerometer also provides a readout of G-forces during reentry.

ALTIMETER

Due to dynamic pressure, static source location, and

FLIGHT DIRECTOR ATTITUDE INDICATOR

ROLL
+ANGULAR VELOCITY-
+ATTITUDE ERROR -

ROLL INDEX

PITCH & YAW
INDEX

ATTITUDE   P   ANGULAR
ERROR   I   VELOCITY
+   T
           C   +
           H

EULER ATTITUDE ON BALL
PITCH - θ = 014°
YAW   - ψ = 034°
ROLL - φ = 330°

ROLL TOTAL
ATTITUDE SCALE

NOTE:
  ALL POLARITIES INDICATE
  VEHICLE DYNAMICS

YAW
+ ATTITUDE ERROR -
+ ANGULAR VELOCITY-

Figure 3-2

instrument error the altimeter (B-51, figure 3-1) is not considered to be an accurate instrument during the launch phase.

The primary function of the altimeter is to provide an adjustable reference (set for barometric pressure on launch date) for parachute deployment for pad/near pad LES aborts. However, the aerodynamic shape of the CM coupled with the static source location produces errors up to 1800 feet. Therefore, the main parachutes must be deployed at an indicated 3800 feet (depends on launch day setting) to ensure deployment above 2000 feet true altitude.

EVENT TIMER

The event timer (P-51, figure 3-1) is a critical display because it is the primary cue for the transition of abort modes, manual sequenced events, monitoring roll and pitch program, staging, and S-IVB insertion cutoff. The event timer is started by the liftoff command which enables automatic aborts. The command pilot should be prepared to manually back up its start to ensure timer operation.

The event timer is reset to zero automatically with abort initiation.

MASTER ALARM LIGHT

There are three MASTER ALARM lights, one on main display panel 1 (L-41, figure 3-1), one on main display panel 3 (O-110) and one in the lower equipment bay. The three MASTER ALARM lights ON alert the flight crew to critical spacecraft failures or out-of-tolerance conditions identified in the caution and warning light array. After extinguishing the alarm lights, action should be initiated to correct the failed or out-of-tolerance subsystem. If crew remedial action does not correct the affected subsystem, then an abort decision must be made based on other contingencies. Secondary abort cues will come from subsystem displays, ground verification, and physiological indications.

NOTE

The Commander's MASTER ALARM light (L-41) will not illuminate during the launch phase, but the other two MASTER ALARM lights can illuminate and the alarm tone will sound.

LV TANK PRESS GAUGES

The LV TANK PRESS gauges (X-46, figure 3-1) indicate ullage pressures in the S-II and S-IVB oxidizer tank. Prior to S-II/S-IVB separation the two left-hand pointers indicate the pressure in the S-II fuel tank. Subsequent to S-II/S-IVB separation these same two pointers indicate S-IVB oxidizer pressure. The two right-hand pointers indicate S-IVB fuel tank pressure until LV/spacecraft separation.

LAUNCH VEHICLE NORMAL CONTROLS

GUIDANCE SWITCH

The GUIDANCE switch is a two position guarded toggle switch with the two positions being IU and CMC (AD-57, figure 3-1). The switch controls a relay in the IU which selects either the IU or the CMC in the spacecraft as the source of flight control attitude error signals for the LV. The normal position of the GUIDANCE switch is IU. Placing the switch in the CMC position permits spacecraft control of the LV under certain conditions.

Guidance Reference Failure Condition

During the LV burn modes, when the LVDC recognizes a guidance reference failure and turns on the LV GUID light, the GUIDANCE switch can be placed in the CMC position. (The switch function is interlocked in the LVDC such that a guidance reference failure must be recognized before the CMC switch position will be honored.) With the switch in the CMC position and LVDC recognition of a guidance reference failure, the LV will receive attitude error signals from the CMC via the LVDC. During the S-IC burn phase the CMC provides attitude error signals based on preprogrammed polynomial data. During S-II and S-IVB burn phases the rotational hand control (RHC) is used to generate the attitude error signals. The amount of attitude error transmitted to the LVDC is a function of how long the RHC is out of detent. The RHC in this mode is not a proportional control.

Timebase 6 will not initiate automatically if a guidance reference failure has occurred. In this circumstance, and only this circumstance, the astronaut can initiate time base 6 by insertion of a command through the DSKY which will supply an S-IVB Ignition Sequence Start signal to the IU. Also, this is the only way in which T_6 can be initiated following a guidance reference failure. Receipt of this S-IVB Ignition Sequence Start signal in the IU after T_5 + 100 seconds will start T_6 immediately. This signal must remain on for at least 5 seconds. Since the Translunar Injection Inhibit signal is disabled by the S-IVB Ignition Sequence Start signal for as long as the discrete is present in the IU, the discrete should not remain on past T_6 + 41.0 seconds.

No Guidance Reference Failure Condition

During the coast modes, T_5 and T_7, the guidance reference failure is not interlocked with the GUIDANCE switch and the spacecraft can assume control of the LV any time the switch is placed in the CMC position. With the GUIDANCE switch in the CMC position and no guidance reference failure, the LVDC will function in a follow-up mode. When LV control is returned to the LVDC during T_5, attitude orientation will be maintained with reference to local horizontal. When LV control is returned to the LVDC in T_7, attitude orientation will be maintained inertially. The LVDC will maintain the attitude existing at the time control is returned from the spacecraft until the next programmed maneuver, at which time the LVDC returns to its programmed attitude timeline.

NOTE

If spacecraft takeover has been effected in T_5, LV control will automatically be returned to the LVDC at the start of T_6.

XLUNAR SWITCH

The XLUNAR switch is a two position toggle switch with the two positions being INJECT and SAFE (AD-59, figure 3-1). Normal position of the switch is INJECT. In this position the LVDC will sequence the S-IVB systems through the steps necessary to accomplish a restart and inject the spacecraft into the mission trajectory.

During certain periods of time the switch can be placed in the SAFE position to inhibit restart of the S-IVB. This translunar injection (TLI) inhibit is a temporary action the first time it is used but if a second injection opportunity is inhibited it becomes final and no restart of the S-IVB can be subsequently accomplished. Timing of these options is under control of the LVDC and is as follows:

1. If the XLUNAR switch is placed in the SAFE position prior to the start of T_6 the LVDC will accept the signal upon starting T_6 and inhibit the restart sequence.

2. If the XLUNAR switch is placed in the SAFE position after T_6 is started, but prior to T_6 + 41.0 seconds, the LVDC will accept the signal at T_6 + 41.0 seconds and inhibit O_2H_2 burner ignition and the remainder of the restart sequence.

3. If the XLUNAR switch is placed in the SAFE position after O_2H_2 burner ignition (T_6 + 41.3 seconds), but before T_6 + 5 minutes 41.3 seconds, the LVDC will accept the signal at T_6 + 5 minutes 41.3 seconds and inhibit ambient repressurization.

4. If the XLUNAR switch is placed in the SAFE position after ambient repressurization (T_6 + 8 minutes 17.3 seconds), but before TLI commit at T_6 + 9 minutes 20 seconds, the LVDC will accept the signal within two seconds and inhibit S-IVB ignition.

5. If the XLUNAR switch is placed in the SAFE position after TLI commit the LVDC will not accept the signal and S-IVB restart will occur.

WARNING

After TLI commit (T_6 + 9 minutes 20 seconds) the only way remaining to effect TLI inhibit is with the T-handle.

Refer to OPERATION SEQUENCE paragraph of section VII for definition of time bases.

EDS POWER SWITCH

The EDS POWER switch should be in the EDS POWER position during prelaunch and launch operations. The switch, if placed in the OFF position results in an "EDS Unsafe" function. This function has the following effect on the launch countdown.

1. If the switch is turned OFF before automatic sequence, the automatic sequence will not be entered.

2. If the switch is turned OFF during the automatic sequence (starting at T-187 seconds), but prior to T-30.0 seconds, the countdown is stopped at T-30.0 seconds and the countdown is recycled to T-24 minutes.

3. If the switch is turned OFF after T-30 seconds, but prior to T-16.2 seconds, the countdown is stopped at T-16.2 seconds and the countdown is recycled to T-24 minutes.

4. If the switch is turned OFF after T-16.2 seconds, but prior to T-8.9 seconds (ignition command), the countdown is stopped at T-8.9 seconds and the launch is recycled to T-24 minutes (or is scrubbed for that day).

5. If the switch is turned OFF after T-8.9 seconds, but prior to T-50 ms (launch commit), the countdown is stopped immediately and the launch is scrubbed for that day.

6. After T-50 ms, the switch will not stop launch; however, the S C EDS power off will have the following effect on the mission.

   a. The EDS displays will not be operative.

   b. The Auto-Abort capability will not be enabled; however, manual abort can be initiated.

WARNING

If the SC EDS POWER switch is returned to the ON position after liftoff, an immediate abort may result, depending upon which relays in the EDS circuit activate first.

UP TLM SWITCHES

The UP TLM switches are a pair of two position toggle switches, one for the CM and one for the IU, both having ACCEPT and BLOCK positions (0-66, figure 3-1). Their function is to control relays in the CM and IU which either block or pass ground commands from the Command Communications System (CCS) to the CMC or LVDC. Further discussion will be limited to the IU switch. Throughout LV burn phases the IU switch is flight program inhibited and has no effect if used. The following table shows the time periods when the IU switch is effective for various ground commands if placed in the ACCEPT position.

COMMAND MESSAGE	TIME PERIOD
Maneuver Update Maneuver Inhibit Time Base Update	from T_5 + 0 to T_6-9 seconds and from T_7+0 to end of S-IVB operations
Navigation Update Sector Dump and Single Memory Location Dump	from T_5 + 100 seconds to T_6-9 seconds and from T_7 + 20 seconds to end of S-IVB operations
Generalized Switch Selector Command	from T_5 + 0 to T_6 + 9 minutes 20 seconds and from T_7 + 0 to end of S-IVB operations

T_8 Enable (S-IVB evasive maneuver & lox dump)	from T_7 + 2 hours to end of S-IVB operations
Target Update	from T_5 + 0 to T_6 -9 seconds
Execute Maneuver A (Communications Attitude)	from T_7 + 0 to end of S-IVB operations
TD&E Enable (Alternate Mission in Earth Orbit)	from T_5 + 0 to T_6 -9 seconds

In T_5 and T_7, when the switch is in the IU ACCEPT position (command window open), a ground command can be sent which will lock the relay in the IU ACCEPT position. At approximately T_7 + 20 minutes the flight program will issue a switch selector command which locks the relay in the IU ACCEPT position.

NOTE

During T_5 if spacecraft separation is required, the IU switch should be placed in the ACCEPT position prior to separation to allow a ground command to lock the relay in the ACCEPT position. This allows for safing of the S-IVB stage in earth orbit by ground command.

S-IVB/LM SWITCH

The S-IVB/LM switch is a two position guarded toggle switch, the two positions being OFF and SEP (Y-63, figure 3-1). Its purpose is to provide a means of manually initiating separation of the spacecraft from the S-IVB stage following CSM separation, transposition and docking.

When the switch is placed in the SEP position, ordnance charges are fired which separate the tension ties securing the LM legs to the spacecraft LM adapter (SLA). Two 30 millisecond parallel time delays are also started at the same time. When either of these time delays expires the LM tension tie firing circuit is deadfaced and a guillotine is fired severing the LV/SC umbilical. This leaves the spacecraft (CSM and LM) ready for extraction from the S-IVB/IU/SLA.

LAUNCH VEHICLE EDS CONTROLS

EDS SWITCH

The EDS switch is a two position toggle switch with the two positions being AUTO and OFF (Y-58, figure 3-1). Prior to liftoff the EDS switch is placed in the AUTO position so that an automatic abort will be initiated if:

1. a LV structural failure occurs between the IU and the CSM,

2. two or more S-IC engines drop below 90% of rated thrust,

3. LV rates exceed 4 degrees per second in pitch or yaw or 20 degrees per second in roll.

The two engine out and LV rate portions of the auto abort system can be manually disabled, individually, by the crew. However, they are automatically disabled by the LV sequencer prior to center engine cutoff.

CSM/LM FINAL SEP SWITCHES

The CSM/LM FINAL SEP switches are a pair of two position guarded toggle switches (Y-60, figure 3-1). Their purpose is to provide a means of manually initiating final separation of the LM docking ring. During a normal entry or an SPS abort, the docking ring must be jettisoned by actuation of the CSM/LM FINAL SEP switches. Failure to jettison the ring could possibly hamper normal earth landing system (ELS) functions.

CM/SM SEP SWITCHES

The two CM/SM SEP switches (Y-62, figure 3-1) are redundant, momentary ON, guarded switches, spring loaded to the OFF position. They are normally used by the Command Pilot to accomplish CM/SM separation prior to the reentry phase. These switches can also be used to initiate an LES abort in case of a failure in either the EDS or the translational controller. All normal post-abort events will then proceed automatically. However, the CANARD DEPLOY pushbutton (Y-50, figure 3-1) should be depressed 11 seconds after abort initiation, because canard deployment and subsequent events will not occur if the failure was in the EDS instead of the translational controller. If the CANARD DEPLOY pushbutton is depressed, all automatic functions from that point on will proceed normally.

PRPLNT SWITCH

The PRPLNT switch is a two position toggle switch with the two positions being DUMP AUTO and RCS CMD (AA-57, figure 3-1). The switch is normally in the DUMP AUTO position prior to liftoff in order to automatically dump the CM reaction control system (RCS) propellants, and fire the pitch control (PC) motor if an abort is initiated during the first 42 seconds of the mission. The propellant dump and PC motor are inhibited by the SC sequencer at 42 seconds. The switch in the RCS CMD position will inhibit propellant dump and PC motor firing at any time.

ABORT SYSTEM - 2 ENG OUT SWITCH

The 2 ENG OUT switch is a two position toggle switch, the two positions being AUTO and OFF (AA-58, figure 3-1). The purpose of this switch is to enable or disable EDS automatic abort capability for a two engine out condition. Normal position of the switch is AUTO, which enables the EDS automatic abort capability. With the switch in OFF the EDS automatic abort capability is disabled.

ABORT SYSTEM - LV RATES SWITCH

The LV RATES switch is a two position toggle switch, the two positions being AUTO and OFF (AA-59, figure 3-1). The purpose of this switch is to enable or disable EDS automatic abort capability for excessive LV rates. Normal position of the switch is AUTO, which enables the EDS automatic abort capability. With the switch in OFF the EDS automatic abort capability is disabled.

ABORT SYSTEM - TWR JETT SWITCHES

There are two redundant TWR JETT guarded toggle switches (AA-62, figure 3-1). When these switches are placed in AUTO, explosive bolts and the tower jettison motor are fired to jettison the LET. Appropriate relays are also de-energized

so that if an abort is commanded, the SPS abort sequence and not the LES sequence will occur.

S-II/S-IVB SWITCH

The S-II/S-IVB switch is a two position guarded toggle switch with positions of LV STAGE and OFF (AD-58, figure 3-1). This switch performs a dual function. In the LV STAGE position it is used to manually initiate the staging sequence of the S-II from the S-IVB and/or manually initiate S-IVB engine cutoff. The switch controls relays in the IU to accomplish these functions.

Subsequent to S-II engines start the S-II/S-IVB separation sequence can be manually initiated by placing the S-II/S-IVB switch in the LV STAGE position.

WARNING

If the switch is used for this purpose it must be reset to the OFF position to enable its S-IVB cutoff function.

Subsequent to S-IVB engine start, in either first burn or second burn, the S-IVB cutoff sequence can be manually initiated by placing the S-II/S-IVB switch in the LV STAGE position. This signal, from the S-II/S-IVB switch, is examined by the LVDC on a nominal computer cycle basis and there may be a delay of up to two seconds between switch positioning and initiation of the cutoff sequence.

MAIN RELEASE SWITCH

The MAIN RELEASE switch (AG-57, figure 3-1) is a toggle switch guarded to the down position. It is moved to the up position to manually release the main chutes after the Command Module has landed. No automatic backup is provided. This switch is armed by the ELS LOGIC switch ON and the 10K barometric switches closed (below 10,000 feet altitude).

NOTE

The ELS AUTO switch must be in the AUTO position to allow the 14 second timer to expire before the MAIN CHUTE RELEASE switch will operate.

ELS SWITCHES

There are two ELS two position toggle switches (AC-48, figure 3-1). The left hand switch is guarded to the OFF position and should only be placed in the LOGIC position during normal reentry or following an SPS abort, and then only below 45,000 feet altitude. If the switch is activated at any time below 30,000 feet (pressure altitude), the landing sequence will commence, i.e., LES and apex cover jettison and drogue deployment. If activated below 10,000 feet altitude, the main chutes will also deploy.

ELS LOGIC is automatically enabled following any manual or auto EDS initiated LES abort. It should be manually backed up if time permits.

WARNING

Do not use ELS LOGIC switch during normal launch. Activation of ELS LOGIC switch below 10,000 feet altitude will initiate the landing sequence causing LES and apex cover jettison, and deployment of drogue and main chutes simultaneously.

The right hand switch is not guarded and has positions of AUTO and MAN. Its normal position is AUTO to enable the automatic sequencing of the ELS during a CM descent period. If the switch is placed in MAN it will inhibit all automatic sequencing of the ELS.

CM RCS PRESS SWITCH

The CM RCS PRESS switch is a two position guarded toggle switch (0-68, figure 3-1). Any time the CM is to be separated from the SM, the CM RCS must be pressurized. The normal sequence of events for an abort or normal CM/SM SEP is to automatically deadface the umbilicals, pressurize the CM RCS, and then separate the CM/SM. However, if the automatic pressurization fails, the CM RCS can be pressurized by the use of the CM RCS PRESS switch.

ABORT CONTROLS

TRANSLATIONAL CONTROLLER

The TRANSLATIONAL CONTROLLER, which is mounted on the left arm of the commanders couch, can be used to accomplish several functions. A manual LES abort sequence is initiated by rotating the T-handle fully CCW. This sends redundant engine cutoff commands to the LV (engine cutoff from the SC is inhibited during the first 30 seconds of flight), initiates CM/SM separation, fires the LES motors, resets the SC sequencer and initiates the post abort sequence. For a manually initiated SPS abort, the CCW rotation of the T-handle commands LV engine cutoff, resets the SC sequencer and initiates the CSM/LV separation sequence.

NOTE

Returning the T-handle to neutral before the 3 seconds expires results only in an engines cutoff signal rather than a full abort sequence.

CW rotation of the T-handle transfers control of the SC from the CMC to the SCS.

The T-handle can also provide translation control of the CSM along one or more axes. The T-handle is mounted approximately parallel to the SC axis; therefore, T-handle movement will cause corresponding SC translation. Translation in the + X axis can also be accomplished by use of the direct ullage pushbutton; however, rate damping is not available when using this method.

SEQUENCER EVENTS · MANUAL PUSHBUTTONS

These are a group of covered pushbuttons switches (X-51,

figure 3-1) which provide a means of manual backup for abort and normal reentry events which are otherwise sequenced automatically.

LES MOTOR FIRE Switch

The LES MOTOR FIRE switch is used to fire the launch escape motor for an LES abort if the motor does not fire automatically. It is also a backup switch to fire the LET jettison motor in the event the TWR JETT switches fail to ignite the motor.

CANARD DEPLOY Switch

The CANARD DEPLOY switch is used to deploy the canard in the event it does not deploy automatically during an abort.

CSM/LV SEP Switch

The CSM/LV SEP switch is used as the primary means of initiating CSM/LV separation after the ascent phase of the mission. When the switch is pressed it initiates ordnance devices which explosively sever the SLA, circumferentially around the forward end, and longitudinally, into four panels. The four panels are then rotated away from the LM by ordnance thrusters. Upon reaching an angle of 45 degrees spring thrusters jettison the panels away from the SC. The same ordnance train separates the CSM/LV umbilical and the LM/GSE umbilical.

The CSM/LV SEP switch is also used as a backup to initiate separation of the SLA when an SPS abort cannot be initiated from the TRANSLATIONAL CONTROLLER. SPS ullage and firing would have to be manually initiated under these circumstances.

APEX COVER JETT Switch

The APEX COVER JETT switch is used to jettison the APEX COVER in the event it fails to jettison automatically during an abort or a normal reentry.

DROGUE DEPLOY Switch

The DROGUE DEPLOY switch is used to deploy the drogue parachutes in the event they fail to deploy automatically 2 seconds after the 24,000 foot barometric pressure switches close.

MAIN DEPLOY Switch

The MAIN DEPLOY switch is used to deploy main parachutes in the event they fail to deploy automatically when the 10,000 foot barometric pressure switches close. This switch can also be used to manually deploy the main parachutes during mode IA aborts.

CM RCS He DUMP Switch

The CM RCS He DUMP switch is used to initiate depletion of the CM He supply by purging the CM RCS lines and manifolds through the RCS engines if depletion does not occur normally as an automatic function during abort or reentry.

SERVICE PROPULSION SYSTEM (SPS) CONTROL

The SPS provides primary thrust for major velocity changes

subsequent to SC/LV separation and prior to CM/SM separation. The SPS is also used to accomplish mode II, III and IV aborts. SPS ignition can be commanded under control of the CMC, the SCS or manually.

SPS Engine Start

Ullage must be provided prior to any SPS burn. Ullage is normally provided by the THC (+X translation) and backup is by the DIRECT ULLAGE pushbutton (Z-38, figure 3-1). The DIRECT ULLAGE pushbutton is a momentary switch and must be held depressed until the ullage maneuver is complete. It does not provide rate damping. For all modes ΔV THRUST switches (Z-41) will be in the NORMAL position and flight combustion stability monitor (FCSM) SPS A and SPS B switches (AE-47) will be in the RESET/OVERRIDE position. The SPS THRUST light (K-49) will illuminate when the engine is firing.

In the CMC mode THRUST ON is commanded as a result of internal computations. Prerequisites are +X translation (from THC or DIRECT ULLAGE pushbutton), SC CONT switch (W-38) in CMC position and THC in neutral (except for +X translation).

In the SCS mode, the SC CONT switch must be in the SCS position, or the THC rotated CW. SPS ignition is commanded by pressing the THRUST ON pushbutton (AA-38). Prerequisites are +X translation (from THC or DIRECT ULLAGE pushbutton) and EMS/ΔV counter > 0 (L-48). Once SPS ignition has occured in this mode, +X translation and THRUST ON commands can be removed and ignition is maintained until a THRUST OFF command is generated.

In the manual mode the SPS THRUST switch (Z-36) is placed in DIRECT ON. Ignition is maintained until a THRUST OFF command is generated.

WARNING

The SPS THRUST switch is a single-point failure when in the DIRECT ON position with the ΔV THRUST switches in the NORMAL position.

SPS Engine Shutdown

In the CMC mode normal engine shutdown is commanded by the CMC as a result of internal computations.

In the SPS mode engine shutdown can be commanded by:

1. FCSM automatically if instability is detected.

2. EMS ΔV counter running down to O.

3. placing the ΔV THRUST switches (both) to OFF

In the manual mode the SPS THRUST switch must be placed to NORMAL or the ΔV THRUST switches (both) to OFF.

WARNING

Both ΔV THRUST switches must be placed in the OFF position if they are used to stop the engine.

Thrust Vector Control

Four gimbal motors control the SPS engine position in the pitch and yaw planes; two motors for each plane. These motors are activated by the SPS GIMBAL MOTORS switches (AC-42, figure 3-1).

NOTE

The motors should be activated one at a time due to high current drain during the start process.

Control signals to the gimbal motors can come from the CMC, SCS or the RHC. Gimbal trim thumbwheels (Z-46) can also be used to position the gimbals in the SCS ΔV mode.

The TVC GIMBAL DRIVE switches (AE-42) are three position toggle switches. Their purpose is to select the source and routing of TVC signals. The switches are normally in the AUTO position.

STABILITY CONTROL SYSTEM (SCS)

The SCS is a backup system to the primary guidance navigation and control system (PGNCS). It has the capability of controlling rotation, translation, SPS thrust vector and associated displays. Switches which affect the SCS are discussed in the following paragraphs.

AUTO RCS SELECT Switches

Power to the RCS control box assembly is controlled by 16 switches (O-14, figure 3-1). Individual engines may be enabled or disabled as required. Power to the attitude control logic is also controlled in this manner, which thereby controls all attitude hold and/or maneuvering capability using SCS electronics (automatic coils). The direct solenoids are not affected as all SCS electronics are bypassed by activation of the DIRECT RCS switches.

NOTE

The automatic coils cannot be activated until the RCS ENABLE is activated either by the MESC or manually.

DIRECT Switches

Two DIRECT switches (W-35, figure 3-1) provide for manual control of the SM RCS engines. Switch 1 controls power to the direct solenoid switches in rotational controller 1 and switch 2 controls power to the direct solenoid switches in rotational controller 2. In the down position switch 1 receives power from MNA and switch 2 receives power from MNB. In the up position both switches receive power from both MNA and MNB. Manual control is achieved by positioning the rotational control hardover to engage the direct solenoids for the desired axis change.

ATT SET Switch

The ATT SET switch (R-40, figure 3-1) selects the source of

total attitude for the ATT SET resolvers as outlined below.

Position	Function	
UP	IMU	Applies inertial measurement unit (IMU) gimbal resolver signal to ATT SET resolvers. FDAI error needles display differences. Needles are zeroed by maneuvering SC or by moving the ATT SET dials.
DOWN	GDC	Applies GDC resolver signal to ATT SET resolvers. FDAI error needles display differences resolved into body coordinates. Needles zeroed by moving SC or ATT SET dials. New attitude reference is established by depressing GDC ALIGN button. This will cause GDC to drive to null the error; hence, the GDC and ball go to ATT SET dial value.

MANUAL ATTITUDE Switches

The three MANUAL ATTITUDE switches (T-34, figure 3-1) are only operative when the SC is in the SCS mode of operation.

Position	Description
ACCEL CMD	Provides direct RCS firing as a result of moving the rotational controller out of detent (2.5 degrees) to apply direct inputs to the solenoid driver amplifiers.
RATE CMD	Provides proportional rate command from rotational controller with inputs from the BMAG's in a rate configuration.
MIN IMP	Provides minimum impulse capability through the rotational controller.

LIMIT CYCLE Switch

The LIMIT CYCLE switch (T-37, figure 3-1), when placed in the LIMIT CYCLE position, inserts a psuedo-rate function which provides the capability of maintaining low SC rates while holding the SC attitude within the selected deadband limits (limit cycling). This is accomplished by pulse-width modulation of the switching amplifier outputs. Instead of driving the SC from limit-to-limit with high rates by firing the RCS engines all the time, the engines are fired in spurts proportional in length and repetition rate to the switching amplifier outputs.

Extremely small attitude corrections could be commanded which would cause the pulse-width of the resulting output command to be of too short a duration to activate the RCS solenoids. A one-shot multivibrator is connected in parallel to ensure a long enough pulse to fire the engines.

RATE and ATT DEADBAND Switch

The switching amplifier deadband can be interpreted as a rate or an attitude (minimum) deadband. The deadband limits are a function of the RATE switch (T-39, figure 3-1). An additional deadband can be enabled in the attitude control loop with the ATT DEADBAND switch (T-38, figure 3-1) See figure 3-3 for relative rates.

The rate commanded by a constant stick deflection (proportional rate mode only) is a function of the RATE switch position. The rates commanded at maximum stick deflection (soft stop) are shown in figure 3-4.

SC CONT Switch

The SC CONT switch (W-38, figure 3-1) selects the spacecraft control as listed below:

Position	Description
CMS	Selects the G&N system computer controlled SC attitude and TVC through the digital auto-pilot. An auto-pilot control discrete is also applied to CMC.
SCS	The SCS system controls the SC attitude and TVC.

ATTITUDE DEADBAND SWITCH POSITION			
RATE SWITCH POSITION	RATE DEADBAND °/SEC	ATT DEADBAND SWITCH POSITION	
		MINIMUM	MAXIMUM
LOW	±0.2	±0.2°	±4.2°
HIGH	±2.0	±2.0°	±8.0°

Figure 3-3

MAXIMUM PROPORTIONAL RATE COMMAND		
RATE SWITCH POSITION	MAXIMUM PROPORTIONAL RATE COMMAND	
	PITCH AND YAW	ROLL
LOW	0.65°/sec	0.65°/sec
HIGH	7.0 °/sec	20.0 °/sec

Figure 3-4

BMAG MODE Switches

The BMAG MODE switches (Y-33, figure 3-1) select displays for the FDAI using SCS inputs

Consequently, the yaw rate gyro generates an undesirable signal. By coupling a component of the roll signal into the yaw channel, the undesirable signal is cancelled. The .05 G switch performs this coupling function.

EMS FUNCTION Switch

The EMS FUNCTION switch (G-45, figure 3-1) is a 12 position mode selector switch, used as outlined in figure 3-5.

EMS MODE Switch

The EMS MODE switch (I-44, figure 3-1) performs the following functions in the positions indicated:

NORMAL

1. Acts as backup display for G&N entry.

2. Initiates the function selected by the EMS FUNCTION switch.

STBY

1. Resets circuits following tests.

2. Removes power if EMS FUNCTION switch is OFF.

MAN

1. Position for manual entry and "VC modes, or auto entry backup display.

2. Does not permit negative acceleration spikes into countdown circuits.

Threshold Indicator (.05 G Light)

The threshold indicator (.05G light) (K-48, figure 3-1) provides the first visual indication of total acceleration sensed at the reentry threshold (approximately 290,000 feet). Accelerometer output is fed to a comparison network and will illuminate the .05 G lamp when the acceleration reaches .05 G. The light will come on not less than 0.5 seconds or more than 1.5 seconds after the acceleration reaches .05 G and turns off when it falls below .02G (skipout).

Corridor Indicators

By sensing the total acceleration buildup over a given period of time, the reentry flight path angle can be evaluated. This data is essential to determine whether or not the entry angle is steep enough to prevent superorbital "skipout."

The two corridor indicator lights (K-45 and M-45, figure 3-1) are located on the face of the roll stability indicator (L-45).

If the acceleration level is greater than 0.2 G at the end of a ten second period after threshold (.05 G light ON), the upper light will be illuminated. It remains ON until the G-level reaches 2 G's and then goes OFF. The lower light illuminates if the acceleration is equal to or less than 0.2 G at the end of a ten second period after threshold. This indicates a shallow entry angle and that the lift vector should be down for controlled entry, i.e., skipout will occur.

Roll Stability Indicator

The roll stability indicator (L-45, figure 3-1) provides a visual indication of the roll attitude of the CM about the stability axis. Each revolution of the indicator represents 360 degrees of vehicle rotation. The display is capable of continuous rotation in either direction. The pointer up position (0 degrees) indicates maximum lift-up vector (positive lift) and pointer down (180 degrees) indicates maximum lift-down vector (negative lift).

G-V Plotter

The G-V plotter assembly (H-48, figure 3-1) consists of a scroll of mylar tape and a G-indicating stylus. The tape is driven from right to left by pulses which are proportional to the acceleration along the velocity vector. The stylus which scribes a coating on the back of the mylar scroll, is driven in the vertical direction in proportion to the total acceleration.

The front surface of the mylar scroll is imprinted with patterns consisting of "high G-rays" and "exit rays." The "high-G-rays" must be monitored from initial entry velocity down to 4000 feet per second. The "exit rays" are significant only between the entry velocity and circular orbit velocity and are, therefore, only displayed on that portion of the pattern.

The imprinted "high-G-rays" and "exit rays" enable detection of primary guidance failures of the type that would result in either atmospheric exits at supercircular speeds or excessive load factors at any speed. The slope of the G-V trace is visually compared with these rays. If the trace becomes tangent to any of these rays, it indicates a guidance malfunction and the need for manual takeover.

ΔV/RANGE Display

The ΔV/RANGE display provides a readout of inertial flight path distance in nautical miles to predicted splashdown after .05G. The predicted range will be obtained from the PGNCS or ground stations and inserted into the range display during EMS range set prior to entry. The range display will also indicate ΔV (ft/sec) during SPS thrusting.

ABORT MODES AND LIMITS

The abort modes and limits listed in figures 3-6 and 3-7 are based on a nominal launch trajectory. The nominal launch phase voice callouts are listed in figure 3-8.

NOTE

More specific times can be obtained from current mission documentation.

EMERGENCY MODES

Aborts performed during the ascent phase of the mission will be performed by using either the Launch Escape System or the Service Propulsion System.

LAUNCH ESCAPE SYSTEM

The Launch Escape System (LES) consists of a solid propellant launch escape (LE) motor used to propel the CM a safe distance from the launch vehicle, a tower jettison motor,

APOLLO ABORT MODES			
PERIOD	MODE	DESCRIPTION	NOTE
Pad to 42 sec	Mode IA	LET low alt	
42 sec to 100,000 feet (1 min 56 sec)	Mode IB	LET med alt	
100,000 feet to LET JETT (1 min 56 sec) (3 min 16 sec)	Mode IC	LET high alt	
LET JETT to RFL=3200 NM (3 min 16 sec) (10 min 08 sec)	Mode II	Full-lift	(1)
RFL=3200 NM to RHL=3350 NM (10 min 08 sec) (10 min 32 sec)	Mode III CSM NO GO/SLV lofted	Half-lift	(2)
RHL=3350 NM to insertion (10 min 32 sec) (11 min 39 sec)	Mode III CSM NO GO/SLV lofted	SPS retro Half-lift	
COI capability to insertion (9 min 00 sec) (11 min 39 sec)	Mode IV CSM GO	SPS to Orbit	(3)

NOTES
(1) RFL = Full-lift landing range (DSKY: ΔR = -400 NM)
(2) RHL = Half-lift landing range (DSKY: ΔR = 0 NM)
(3) For positive h and S-IVB cutoff beyond the 5 min to apogee line (crew chart) an apogee kick maneuver would be recommended over the Mode IV. Single burn to orbit occurs at 10:16.

Figure 3-6

and a canard subsystem. A complete description on use of the system can be found in the specific mission Abort Summary Document (ASD). A brief description is as follows:

Mode 1A Low Altitude Mode

In Mode IA, a pitch control (PC) motor is mounted normal to the LE motor to propel the vehicle downrange to ensure water landing and escape the "fireball." The CM RCS propellants are dumped through the aft heat shield during this mode to preclude damage to the main parachutes.

The automatic sequence of major events from abort initiation is as follows:

Time	Event
00:00	Abort
	Ox rapid dump
	LE and PC motor fire
00:05	Fuel rapid dump
00:11	Canards deploy
00:14	ELS arm
00:14.4	Apex cover jett
00:16	Drogue deploy
00:18	He purge
00:28	Main deploy

The automatic sequence can be prevented, interrupted, or replaced by crew action.

Mode IB Medium Altitude

Mode IB is essentially the same as Mode IA with the exception of deleting the rapid propellant dump and PC motor features. The canard subsystem was designed specifically for this altitude region to initiate a tumble in the

pitch plane. The CM/tower combination CG is located such that the vehicle will stabilize (oscillations of ± 30 degrees) in the blunt-end-forward (BEF) configuration. Upon closure of barometric switches, the tower would be jettisoned and the parachutes automatically deployed.

As in Mode IA, crew intervention can alter the sequence of events if desired.

Mode IC High Altitude

During Mode IC the LV is above the atmosphere. Therefore, the canard subsystem cannot be used to induce a pitch rate to the vehicle. If the LV is stable at abort, the LET is manually jettisoned and the CM oriented to the reentry attitude. This method provides a stable reentry but requires a functioning attitude reference.

With a failed platform the alternate method will be to introduce a five degree per second pitch rate into the system. The CM/tower combination will then stabilize BEF as in Mode IB. The LES would likewise deploy the parachutes at the proper altitudes.

SERVICE PROPULSION SYSTEM

The Service Propulsion System (SPS) aborts utilize the Service Module SPS engine to propel the CSM combination away from the LV, maneuver to a planned landing area, or boost into a contingency orbit. The SPS abort modes are:

Mode II

The SM RCS engines are used to propel the CSM away from the LV unless the vehicle is in danger of exploding or excessive tumble rates are present at LV/CSM separation. In these two cases the SPS engine would be used due to greater

ABORT LIMITS

RATES

1. Pitch and yaw
   L/O to S-IC/S-II staging   4° per second
   (Excluding staging)
   S-IC/S-II staging to S-IVB CO   9° per second

2. Roll
   L/O to S-IVB CO   20° per second

MAX Q REGION

The following limits represent single cues and are restricted to the time
period of 50 seconds to 2 minutes.

1. Angle of Attack (Q α) = 100%
   Note: The Qα abort cue is valid only when preceeded by one of the
         following cues.

2. Attitude error (roll, pitch, or yaw) = 5°

3. Single S-IC control engine failure occurring later than 50 seconds into
   flight.
   Note: For control engine failures prior to 50 seconds, the Qα abort
         limit is not valid as an abort cue.

PLATFORM FAILURE

1. During S-IC powered flight the two cues for platform failure requiring an
   immediate switchover are:

   a. LV GUID LT - ON

   b. LV RATE LT - ON

2. After LV RATE switch deactivation the primary cue is:

   LV GUID LT - ON

   The secondary cues are:

   a. FDAI attitude

   b. LV rates

   c. Ground confirmation

Figure 3-7 (Sheet 1 of 2)

ABORT LIMITS

AUTOMATIC ABORT LIMITS (L/O UNTIL DEACTIVATE AT 2 MIN)

1. Rate   pitch - yaw $4.0° \pm 0.5°$ per second
                 roll $20.0° \pm 0.5°$ per second

2. Any two engines fail

3. CM to IU breakup

ENGINE FAILURE (S-IC)

1. Single engine failure   **Continue mission**

2. Simultaneous loss two or more engines   **Abort**

ENGINE FAILURE (S-II)

1. Single engine failure   **Continue mission**

2. Double engine failures   **Abort if LV control is lost (rates >9°/sec) otherwise continue mission.**

3. Three or more engine failures   **Abort if prior to S-IVB to orbit capability. Upstage immediately if the failures occur after S-IVB to orbit capability is achieved.**

WARNING

Unless advised by ground, do not upstage after level sense arm (approx. 8:15 GET) Inhibit of S-IVB engine start may occur.

ENGINE FAILURE (S-IVB)   Abort (Mode II, III, or IV)

S-II SECOND-PLANE SEPARATION FAILURE   Immediate abort on ground confirmation.

S-IVB DIFFERENTIAL TANK PRESSURE LIMITS

     ΔP (ORBITAL COAST) $LH_2 > LO_2 = 26$ psid
                        $LO_2 > LH_2 = 36$ psid
                        $LO_2 > 50$ psia

Figure 4-7 (Sheet 1 of 2)

Changed 8 October 1969

BOOST TO ORBIT			
PROG TIME	STA	ACTION/ENTRY *REPORT	OPTION/EVENT
-00:09	LCC	IGNITION*	
+00:01	LCC	LIFTOFF*	UMBILICAL DISCONNECT CMC TO P11 DET & MET START
	CDR	CLOCK START*	
00:10	LCC	CLEAR TOWER*	ABOVE LAUNCH TOWER
00:12	CDR	ROLL & PITCH START*	ROLL AND PITCH PROGRAM START
00:30	CDR	ROLL COMPLETE	ROLL PROGRAM COMPLETE
00:42	MCC	MODE IB*	
	CMP	PRPLNT DUMP - RCS CMD	
00:50	CDR	MONITOR α TO T + 2:00	
00:55	CMP	MONITOR CABIN PRESSURE DECREASING	IF NO DECREASE BY 17K' DUMP MANUALLY
01:56	MCC	MODE IC* (BASED ON 100,000')	
02:00	CMP	EDS AUTO - OFF*	NO AUTO ABORT LIGHT-ON
		EDS ENG - OFF	
		EDS RATES - OFF	
	CMP	α/PC - PC	
02:10	MCC	GO/NO GO FOR STAGING*	
	CDR	GO/NO GO FOR STAGING*	SYSTEMS STATUS
02:15	CDR	INBOARD OFF	ENGINE LIGHT NO. 5 - ON LIFTOFF LIGHT - OUT NO AUTO ABORT LIGHT - OUT
02:40	CDR	OUTBOARD OFF	ENG LIGHTS 1 THRU 4 - ON
02:41	CDR	S-IC/S-II STAGING	ENG LIGHTS - OUT
02:42		S-II IGNITION COMMAND	ENG LIGHTS - ON S-II SEP LIGHT - ON
02:44	CDR	S-II 65%*	ENG LIGHTS - OUT
03:11	CDR	INTERSTAGE JET*	S-II SEP LIGHT - OUT*
03:16	CMP	KEY V82E	
	CMP	TWR JETT (2) - ON* (IF TFF > 1+20)	TOWER JETTISON
	MCC	MODE II*	
	CDR	MAN ATT (PITCH) - RATE CMD	
	CMP	KEY PROCEED	
03:21	CDR	GUIDANCE INITIATE*	IGM START
04:00	CDR	REPORT STATUS*	
	MCC	TRAJECTORY STATUS*	
05:00	CDR	REPORT STATUS*	
05:20	MCC	UPSTAGE CAPABILITY TO COI*	
06:00	CDR	REPORT STATUS*	
		GMBL MOT (4) - START - ON	
		CHECK GPI (MOMENTARILY)	INSURE ANGLES CORRECT
06:15	LMP	OMNI ANT-D (AZ < 96°)	
07:00	CDR	REPORT STATUS*▲	
07:40	CDR	INBOARD OFF	
08:00	CDR	REPORT STATUS*	
08:30	MCC	GO/NO GO FOR STAGING*	SYSTEMS STATUS
	CDR	GO/NO GO FOR STAGING	
09:00	MCC	MODE IV (2 BURN)	V1 ≥ 21,800 FPS h ∼ 100 NM
09:11	CDR	S-II OFF	ENG LIGHTS - ON
09:12	CDR	S-II/S-IVB STAGING	ENG LIGHT NO. 1 - OFF
09:15	CDR	S-IVB IGNITION	ENG LIGHT NO. 1 - ON
09:17	CDR	S-IVB 65%	ENG LIGHT NO. 1 - OFF
	CDR	REPORT STATUS*	
10:16	MCC	MODE IV* (1 BURN)	VI ≥ 23,400 FPS h ∼ 100 NM
	CDR	MODE IV*	
10:00	MCC	REPORT STATUS	
11:39	CDR	SECO*	ENG LIGHT NO. 1 - ON MOMENTARILY
		INSURE ORBIT	
		KEY V82E	
11:49	MCC	INSERTION*	

Figure 3-8

ΔV and attitude control authority. When the CSM is a safe distance and stable, the CM is separated from the SM and maneuvered to a reentry attitude. A normal entry procedure is followed from there.

Mode III

The SPS engine is used to slow the CSM combination (retrograde maneuver) so as to land at a predetermined point in the Atlantic Ocean. The length of the SPS burn is dependent upon the time of abort initiation. Upon completion of the retro maneuver, the CM will separate from

the SM, assume the reentry attitude, and follow normal entry procedures.

Mode IV

The SPS engine can be used to make up for a deficiency in insertion velocity up to approximately 3000 feet per second. This is accomplished by holding the CSM in an inertial attitude and applying the needed ΔV with the SPS to acquire the acceptable orbital velocity. If the inertial attitude hold mode is inoperative, the crew can take over manual control and maneuver the vehicle using onboard data.

S-IC STAGE

TABLE OF CONTENTS

INTRODUCTION

The S-IC stage (Figure 4-1) is a cylindrical booster, 138 feet long and 33 feet in diameter, powered by five F-1 rocket engines. The engines use liquid oxygen as the oxidizer and RP-1 as the fuel. The center engine is fixed and the four outboard engines are gimbaled for vehicle attitude control. Stage and engine characteristics are listed in Figure 1-3.

The S-IC stage provides first stage boost of the Saturn V launch vehicle to an altitude of about 200,000 feet (approximately 36 miles), and provides acceleration to increase the vehicle inertial velocity to 9,029 feet per second. It then separates from the S-II stage and falls to earth about 360 nautical miles downrange.

The stage interfaces structurally and electrically with the S-II stage. It also interfaces structurally, electrically, and pneumatically with two umbilical service arms, three tail service masts, and certain electronic systems by antennae. The major systems of the stage are: structures, environmental control, propulsion, flight control, pneumatic controls, propellants, electrical, instrumentation, and ordnance.

STRUCTURE

The S-IC structure design reflects the requirements of F-1 engines, propellants, control, instrumentation and interfacing systems. The structure maintains an ultimate factor of safety of at least 1.40 applied to limit load and a yield factor of safety of 1.10 on limit load. Aluminum alloy is the primary structural material. The major components, shown in figure 4-1, are the forward skirt, oxidizer tank, intertank section, fuel tank, and thrust structure.

FORWARD SKIRT

The aft end of the forward skirt (figure 4-1) is attached to the oxidizer (lox) tank and the forward end interfaces with the S-II stage. The forward skirt has accomodations for the forward umbilical plate, electrical and electronic canisters, and the venting of the lox tank and interstage cavity. The skin panels, fabricated from 7075-T6 aluminum, are stiffened and strengthened by ring frames and stringers.

OXIDIZER TANK

The 345,000 gallon lox tank is the structural link between the forward skirt and the intertank section. The cylindrical tank skin is stiffened by "integrally machined" T stiffeners. Ring baffles (figure 4-1) attached to the skin stiffeners stabilize the tank wall and serve to reduce lox sloshing. A cruciform baffle at the base of the tank serves to reduce both slosh and vortex action. Support for four helium bottles is provided by the ring baffles. The tank is a 2219-T87 aluminum alloy cylinder with ellipsoidal upper and lower bulkheads. The skin thickness is decreased in eight steps from .254 inches at the aft section to .190 inches at the forward section.

INTERTANK SECTION

The intertank structure provides structural continuity between the lox and fuel tanks. This structure provides a lox fill and drain interface to the intertank umbilical. One opening vents the fuel tank. The corrugated skin panels and circumferential ring frames are fabricated from 7075-T6 aluminum.

FUEL TANK

The 216,000 gallon fuel tank (figure 4-1) provides the load carrying structural link between the thrust structure and intertank structure. The tank is cylindrical, with ellipsoidal upper and lower bulkheads. Antislosh ring baffles are located on the inside wall of the tank and antivortex cruciform baffles are located in the lower bulkhead area. Five lox ducts run from the lox tank, through the RP-1 tank, and terminate at the F-1 engines. A lightweight foam filler, which is bonded to the bottom of the tank, acts as an exclusion riser to minimize unusable residual fuel. The 2219-T87 aluminum skin thickness is decreased in four steps from .193 inches at the aft section to .170 inches at the forward section.

THRUST STRUCTURE

The thrust structure assembly (figure 4-1) redistributes locally applied loads of the five F-1 engines into uniform loading about the periphery of the fuel tank. It also provides support for the five F-1 engines, engine accessories, base heat shield, engine fairings and fins, propellant lines, retrorockets, and environmental control ducts. The lower thrust ring has four holddown points which support the fully loaded Saturn/Apollo (over 6,000,000 pounds) and also, as necessary, restrain the vehicle from lifting off at full F-1 engine thrust. The skin segments are fabricated from 7075-T6 aluminum alloy.

The base heat shield is located at the base of the S-IC stage, forward of the engine gimbal plane. The heat shield provides thermal shielding for critical engine components and base region structural components for the duration of the flight. The heat shield panels are constructed of 15-7 PH stainless steel honeycomb, 1.00-inch thick, brazed to .010 inch steel face sheets.

Each outboard F-1 engine is protected from aerodynamic loading by a conically shaped engine fairing (figure 4-1). The fairings also house the retrorockets and the engine actuator

This Space Intentionally Left Blank

S-IC STAGE

FLIGHT TERMINATION RECEIVERS (2)

INSTRUMENTATION

GOX DISTRIBUTOR

HELIUM CYLINDERS (4)

GOX LINE

ANNULAR RING BAFFLES

CRUCIFORM BAFFLE

SUCTION LINE TUNNELS (5)

CENTER ENGINE SUPPORT

LOX SUCTION LINES (5)

FUEL SUCTION LINES

CABLE TUNNEL

UPPER THRUST RING

HEAT SHIELD

LOWER THRUST RING

F-1 ENGINES (5)

INSTRUMENTATION

RETROROCKETS

FLIGHT CONTROL SERVOACTUATOR

FORWARD SKIRT
120.7 IN

OXIDIZER (LOX) TANK
769 IN

INTERTANK SECTION
262.4 IN

FUEL (RP-1) TANK
517 IN

THRUST STRUCTURE
233.7 IN

FIN C

HEAT SHIELD

ENGINE FAIRING AND FIN

Figure 4-1

This Space Intentionally Left Blank

supports. The fairing components are primarily titanium alloy below station 115.5 and aluminum alloy above this station. Four fixed, titanium covered, stabilizing fins augment the stability of the Saturn V vehicle.

ENVIRONMENTAL CONTROL

During launch preparations the environmental control systems (ECS) protect the S-IC stage and stage equipment from temperature extremes, excessive humidity, and hazardous gases. Conditioned air, from the ground support equipment environmental control unit (GSE-ECU), is forced into the forward skirt and thrust structure where it is used as a temperature and humidity control medium. Approximately two hours before the two upper stages are loaded with cryogenic fluids, gaseous nitrogen (GN_2) replaces conditioned air and is introduced into the S-IC as the conditioning medium. The GN_2 flow terminates at umbilical disconnect since the system is not needed in flight.

FORWARD SKIRT COMPARTMENT - ECS

The environmental control system distributes air or GN_2 to 4 electrical/electronic equipment module canisters located in the forward skirt. Onboard probes control the temperature of the flow medium to maintain canister temperature at 80 ($\pm$20) degrees F. Three phases of the conditioning/purge medium flow are provided to compensate for the environmental imbalances generated by ambient air changes, internal heat and lox load chill effects. The first phase supplies cool, conditioned air to the canisters when onboard electrical systems are energized before cryogenic loading. The second phase occurs when relatively warm GN_2 is substituted for the cool air to offset temperature differences caused by the cryogenic loading. The third phase uses a warmer GN_2 flow to offset temperature decreases caused by second stage J-2 engine thrust chamber chilldown. The air or GN_2 is vented from the canisters and overboard through vent openings in the forward skirt of the S-IC stage. A by-product of the use of the inert GN_2 is the reduction of gaseous hydrogen or oxygen concentrations.

THRUST STRUCTURE COMPARTMENT - ECS

The environmental control system discharges air or $GN2$ through 22 orificed duct outlets directly into the upper thrust structure compartment. The GSE-ECU supplies conditioned air at two umbilical couplings during launch preparations. At 15 minutes before cryogenic loading commences, the flow medium is switched to GN_2 and the temperature varied as necessary to maintain the compartment temperature at 80 ($\pm$20) degrees F. The temperature control compensates for temperature variations caused by ambient air change, and chill effects from lox in the suction ducts, prevalves, and inter-connect ducts. The GN_2 prevents the oxygen concentration in the compartment from exceeding 6 percent.

HAZARDOUS GAS DETECTION

The hazardous gas detection system monitors the atmosphere in the forward skirt and the thrust structure compartment of the S-IC (figures 4-2 and 4-3). This system is not redundant; however, large leaks may be detected by propellant pressure indications displayed in the Launch Control Center.

PROPULSION

The F-1 engine is a single start, 1,530,000 pound fixed thrust, calibrated, bipropellant engine which uses liquid oxygen as the oxidizer and RP-1 as the fuel. Engine features include a bell-shaped thrust chamber with a 10:1 expansion ratio, and detachable, conical nozzle extension which increases the thrust chamber expansion ratio to 16:1. The thrust chamber is cooled regeneratively by fuel, and the nozzle extension is

Figure 4-2

Figure 4-3

cooled by gas generator exhaust gases. Liquid oxygen and RP-1 fuel are supplied to the thrust chamber by a single turbopump powered by a gas generator which uses the same propellant combination. RP-1 fuel is also used as the turbopump lubricant and as the working fluid for the engine fluid power system. The four outboard engines are capable of gimbaling and have provisions for supply and return of RP-1 fuel as the working fluid for a thrust vector control system. The engine contains a heat exchanger system to condition engine supplied liquid oxygen and externally supplied helium for stage propellant tank pressurization. An instrumentation system monitors engine performance and operation. External thermal insulation provides an allowable engine environment during flight operation.

ENGINE OPERATING REQUIREMENTS

The engine requires a source of pneumatic pressure, electrical power, and propellants for sustained operation. A ground hydraulic pressure source, an inert thrust chamber prefill solution, gas generator igniters, gas generator exhaust igniters, and hypergolic fluid are required during the engine start sequence. The engine is started by ground support equipment (GSE) and is capable of only one start before reservicing.

PURGE, PREFILL, AND THERMAL CONDITIONING

A gaseous nitrogen purge is applied for thermal conditioning and elimination of explosive hazard under each engine cocoon. Because of the possibility of low temperatures existing in the space between the engine and its cocoon of thermal insulation, heated nitrogen is applied to this area. This purge is manually operated, at the discretion of launch operations, whenever there is a prolonged hold of the countdown with lox onboard and with an ambient temperature below approximately 55 degrees F. In any case, the purge will be turned on five minutes prior to ignition command and continue until umbilical disconnect.

A continuous gaseous nitrogen purge is required to expel propellant leakage from the turbopump lox seal housing and the gas generator actuator housing. The purge pressure also improves the sealing characteristic of the lox seal. The purge is required from the time propellants are loaded and is continuous throughout flight.

A gaseous nitrogen purge is required to prevent contaminants from entering the lox system through the engine lox injector or the gas generator lox injector. The purge system is activated prior to engine operation and is continued until umbilical disconnect.

From T-15 hours to T-13 hours, an ethylene glycol solution fills the thrust chamber tubes and manifolds of all five engines. This inert solution serves to smooth out the combustion sequence at engine start. Flow is terminated by a signal from an observer at the engines. At approximately T-11 minutes, 50 gallons are supplied to top off the system to compensate for liquid loss that occurred during engine gimbaling.

POGO SUPPRESSION SYSTEM

The POGO suppression system (figure 4-4) utilizes the lox prevalve cavities of the four outboard engines as surge chambers to suppress the POGO phenomenon. The lox prevalve cavities are pressurized with gaseous helium (GHe) at

T-11 minutes from ground supply by opening the POGO suppression control valves. During the initial fill period (T-11 to T-9 minutes), the filling of the valves is closely monitored, utilizing measurements supplied by the liquid level resistance thermometers R3 (primary) and R_2 (backup). The GHe ground fill continues to maintain the cavity pressure until umbilical disconnect. Following umbilical disconnect the cavity pressure is maintained by the cold helium spheres located in the lox tank.

Status on system operation is monitored through two pressure transducers and four liquid level resistance thermometers. The pressure transducer (0-800 psia) monitors system input pressure. A second pressure transducer (0-150 psia) monitors the pressure inside the No. 1 engine lox prevalve cavity. These pressure readings are transmitted via telemetry to ground monitors. The liquid level within the prevalves is monitored by four liquid level resistance thermometers in each prevalve. These thermometers transmit a "wet" (colder than -165 degrees centigrade) and a "dry" (warmer than -165 degrees centigrade) reading to ground monitors.

ENGINE SUBSYSTEMS

The subsystems of the F-1 engine shown in figure 4-5 are the turbopump, checkout valve, hypergol manifold, heat exchanger, main fuel valve and main lox valve. Subsystems not shown are the gas generator, 4-way control valve, and pyrotechnic igniters.

Hypergol Manifold

The hypergol manifold consists of a hypergol container, an ignition monitor valve (IMV), and an igniter fuel valve (IFV). The hypergol solution is forced into the thrust chamber by the fuel where combustion is initiated upon mixing with the lox. The IFV prevents thrust chamber ignition until the turbopump pressure has reached 375 psi. The IMV prevents opening of the main fuel valves prior to hypergolic ignition. A positive hypergol cartridge installed indication is provided by sensors and is a prerequisite to the firing command.

Control Valve - 4-Way

The 4-way control valve directs hydraulic fluid to open and close the fuel, lox, and gas generator valves. It consists of a filter manifold, a start and stop solenoid valve, and two check valves.

Turbopump

The turbopump is a combined lox and fuel pump driven through a common shaft by a single gas turbine.

Gas Generator

The gas generator (GG) provides the gases for driving the turbopump. Its power output is controlled by orifices in its propellant feed lines. The gas generator system consists of a dual ball valve, an injector, and a combustor. Combustion is initiated by two pyrotechnic igniters. Total propellant flow rate is approximately 170 lb/sec at a lox/RP-1 mixture ratio of 0.42:1. The dual ball valve must be closed prior to fuel loading and must remain closed to meet an interlock requirement for engine start.

Heat Exchanger

The heat exchanger expands lox and cold helium for propellant tank pressurization. The cold fluids, flowing through separate heating coils, are heated by the turbopump exhaust. The warm expanded gases are then routed from the heating coils to the propellant tanks.

Main Fuel Valve

There are two main fuel valves per engine. They control flow of fuel to the thrust chamber. The main fuel valve is a fast acting, pressure balanced, poppet type, hydraulically operated valve. Movement of the poppet actuates a switch which furnishes valve position signals to the telemetry system. This valve is designed to remain open, at rated engine pressures and flowrates, if the opening control pressure is lost. Both valves must be in the closed position prior to fuel loading or engine start.

Main Lox Valve

The two main lox valves on each engine control flow to the thrust chamber. These valves are fast acting, pressure balanced, poppet type, hydraulically operated valves. A sequence valve operated by the poppet allows opening pressure to be applied to the GG valve only after both main lox valve poppets have moved to a partially open position. This valve is designed to remain open, at rated engine pressures and flowrates, if the opening control pressure is lost. Both main lox valves must be in the closed position prior to lox loading or engine start.

Checkout Valve

The checkout valve directs ground supplied control fluid from the engine back to ground during engine checkout. Approximately 30 seconds prior to the firing command the valve is actuated to the engine position. In this position it directs control fluid to the No. 2 turbopump inlet. An ENGINE POSITION indication is required from this valve prior to, and interlocked with, the forward umbilical disconnect command in order for the terminal countdown to continue.

High Voltage Igniters

Four high voltage igniters, two in the gas generator (GG)

Figure 4-4

body and two in the engine thrust chamber nozzle extension ignite the GG and the fuel rich turbopump exhaust gases. They are ignited during the F-1 engine start sequence by application of a nominal 500 volts to the igniter squibs.

ENGINE OPERATION

Engine operation includes starting, mainstage and cutoff. The starting and cutoff phases are periods of transition in which a sequence of activities occurs. These phases are developed in detail in the following paragraphs.

Engine Start

The engine start and transition to mainstage is illustrated in Figure 4-6 for a typical single engine.

Engine Cutoff

The normal inflight cutoff sequence is center engine first, followed by the four outboard engines. At approximately 2 minutes, 15 seconds, the center engine is programmed by the LVDC for cutoff. This command also initiates time base No.

2 (T_2 + 0.0). The LVDC provides a backup center engine shutoff signal. Outboard engine cutoff is enabled by a signal from the LVDC which permits cutoff to be initiated upon energization of two out of four optical type depletion sensors in either the lox or the fuel tank. (Lox depletion is most probable). The sensors start a timer which, upon expiration, energizes the 4-way control valve stop solenoid on each outboard engine. Time base No. 3 (T_3 + 0.0) is initiated at this point. The remaining shutdown sequence of the outboard engines is the same as for the center engines as explained in Figure 4-7.

Emergency Engine Cutoff

In an emergency, the engine will be cut off by any of the following methods: Ground Support Equipment (GSE) Command Cutoff, Range Safety Command Cutoff, Thrust Not OK Cutoff, Emergency Detection System, Outboard Cutoff System.

GSE has the capability of initiating engine cutoff anytime until umbilical disconnect. Separate command lines are supplied through the aft umbilicals to the engine cutoff relays and prevalve close relays.

Figure 4-5

The Range Safety Command Cutoff provides the capability of engine cutoff any time after liftoff. If it is determined during flight that the vehicle has gone outside the established corridor, the Range Safety Officer will send commands to effect engine cutoff and propellant dispersion.

Three thrust OK pressure switches are located on each F-1 engine thrust chamber fuel manifold and sense main fuel injection pressure. If the pressure level drops below the deactivation level of two of the three pressure switches, an engine cutoff signal is initiated. This signal is inhibited until T_1 + 14.0 seconds to take advantage of any thrust a failing engine might produce until the tower has been cleared. Subsequent to tower clearance, the thrust not OK signal will initiate shutdown of the failing engine. The signal is also transmitted to the IU where it is monitored by the EDS to detect multiple engine failures. However, a timer in the IU inhibits EDS multiple engine cutoff commands until T_1 + 30.0 seconds. This allows the launch vehicle time to clear the launch area and proceed far enough down range to ensure a water landing of the aborted spacecraft. Subsequent to T_1 + 30.0 seconds, the EDS multiple engine cutoff circuitry is enabled until 0.8 seconds before CECO. While this EDS circuitry is enabled, a thrust not OK signal from two or more engines will cause the EDS to initiate an ENGINES OFF command which will shut down all engines on the stage. EDS multiple engine cutoff can be manually inhibited by crew action at any time.

The outboard engine cutoff system is activated 0.1 second before center engine cutoff, and is similar in function to the EDS. This system monitors only the outboard engines and sends cutoff signals to the remaining engines if shutdown of two adjacent engines is detected. The system precludes a stage breakup in the event two adjacent outboard engines fail following CECO.

In summary, emergency engine cutoff is possible under the following conditions:

CONDITION	COMMAND
Before umbilical disconnect	GSE cutoff command
Any time after liftoff	Range Safety Officer cutoff command
Liftoff to T_1 +14.0 sec.	Automatic cutoff inhibited
T_1+14.0 sec to normal engine cutoff	Automatic cutoff command to failing engine only
T_1+30.0 sec. to 0.8 sec. before CECO	Automatic cutoff command to all engines if two or more engines fail
0.1 sec before CECO to OBECO	Automatic cutoff command to all engines if two adjacent outboard engines fail.

FLIGHT CONTROL

The S-IC flight control system gimbals the four outboard engines to provide attitude control during the S-IC burn phase. See Section VII for a detailed discussion of the Saturn V flight control.

FLUID POWER

There are five fluid power systems, one system for each engine, on the S-IC stage. During test, prelaunch checkout and engine start the hydraulic pressure is supplied from a GSE pressure source. After engine start, the hydraulic pressure is generated by the engine turbopump. Pressure from either source is made available to the engine valves, such as the main fuel and lox valves and the igniter fuel valve. These valves are sequenced and controlled by the terminal countdown sequencer, stage switch selector and by mechanical or fluid pressure means as described in the paragraphs on Propulsion and Electrical. The two flight control servoactuators on each outboard engine are also powered by this system as shown in figure 4-8.

The fluid power system uses both RJ-1 ramjet fuel and RP-1 rocket propellant as the hydraulic fluid. The RJ-1 is used by the Hydraulic Supply and Checkout Unit (GSE pressure source). RP-1 is the fuel used in the S-IC stage and as a hydraulic fluid, is pressurized by the engine turbopump. These two hydraulic fluids are separated by check valves and their return flow is directed to GSE or stage by the ground checkout valve. Drilled passages in the hydraulic components (valves and servoactuators) permit a flow of fluid to thermally condition the units and to bleed gases from the fluid power system.

HYDRAULIC SERVOACTUATOR

The servoactuator (figure 4-9) is the power control unit for converting electrical command signals and hydraulic power into mechanical outputs to gimbal the engines on the S-IC stage. The flight control computer (IU) receives inputs from the guidance system in the IU and sends signals to the servoactuators to gimbal the outboard engines in the direction and magnitude required. An integral mechanical feedback varied by piston position modifies the effect of the IU control signal. A built-in potentiometer senses the servoactuator position and transmits this information to the IU for further transmission via telemetry to the ground. On vehicles SA-508 and on, the actuators are equipped with a ±3 degree position indicator to provide actuator position information to ESE for detection of an actuator hard-over failure prior to launch commit.

The servoactuators are mounted 90 degrees apart on each outboard engine and provide for engine gimbaling at a rate of 5 degrees per second and a maximum angle of ±5.0 degrees square pattern.

PNEUMATIC CONTROLS

The pneumatic control system (figure 4-10) provides a pressurized nitrogen supply for command operation of various pneumatic valves. Pneumatic control of the fuel and lox fill and drain valves and the No. 2 lox interconnect valve is provided directly from GSE. Lox interconnect valves No. 1, 3 and 4 are controlled by the onboard pneumatic system.

The pneumatic control system for those valves which must be controlled during flight (fuel and lox prevalves, lox and fuel vent valves) is charged by a GSE nitrogen source at 3200 psi. The system storage bottle is charged through onboard control valves and filters. A pressure regulator reduces the supplied pressure to 750 psi. There are direct lines from the GSE to the prevalve solenoid valves to provide pressure for emergency engine shutdown prior to liftoff. Closing time of the prevalves is controlled by orifices in these lines.

ENGINE START

Figure 4-6 (Sheet 1 of 2)

ENGINE START

1 Engine start is part of the terminal count-down sequence. When this point in the count-down is reached, the ignition sequencer controls starting of all five engines.

2 Checkout valve moves to engine return position.

3 Electrical signal fires igniters (4 each engine).

   a) Gas generator combustor and turbine exhaust igniters burn igniter links to trigger electrical signal to start solenoid of 4-way control valve.

   b) Igniters burn approximately six seconds.

4 Start solenoid of 4-way control valve directs GSE hydraulic pressure to main lox valves.

5 Main lox valves allow lox to flow to thrust chamber and GSE hydraulic pressure to flow through sequence valve to open gas generator ball valve.

6 Propellants, under tank pressure, flow into gas generator combustor.

7 Propellants are ignited by flame of igniters.

8 Combustion gas passes through turbopump, heat exchanger, exhaust manifold and nozzle extension.

9 Fuel rich turbine combustion gas is ignited by flame from igniters.

   a) Ignition of this gas prevents back-firing and burping.

   b) This relatively cool gas (approximately 1,000°F) is the coolant for the nozzle extension.

10 Combustion gas accelerates the turbopump, causing the pump discharge pressure to increase.

11 As fuel pressure increases to approximately 375 psig, it ruptures the hypergol cartridge.

12 The hypergolic fluid and fuel are forced into the thrust chamber where they mix with the lox to cause ignition.

TRANSITION TO MAINSTAGE

13 Ignition causes the combustion zone pressure to increase.

14 As pressure reaches 20 psig, the ignition monitor valve directs fluid pressure to the main fuel valves.

15 Fluid pressure opens main fuel valves.

16 Fuel enters thrust chamber. As pressure increases the transition to mainstage is accomplished.

17 The thrust OK pressure switch (which senses fuel injection pressure) picks up at approximately 1060 psi and provides a THRUST OK signal to the IU.

Figure 4-6 (Sheet 2 of 2)

ENGINE CUTOFF

PROGRAMMED CUTOFF
EDS CUTOFF
GSE CUTOFF
RANGE SAFETY CUTOFF
THRUST O.K. PRESSURE SWITCH DROPOUT (2 OUT OF 3)

OR

THRUST O.K. PRESSURE SWITCH (3)

CHECKOUT VALVE

FROM FUEL TANK   FROM LOX TANK   FROM FUEL TANK

ELECTRICAL
SENSE LINE
STAGE FUEL
LOX

1 The 4-Way Control Valve Stop Solenoid is energized, which routes closing pressure to the following valves.

2 Gas Generator Ball Valve closes.

3 Main Lox Valves (2 ea) close.

4 Main Fuel Valves (2 ea) close.

5 The Thrust Chamber pressure decay causes the thrust OK pressure switch to drop out (3 ea).

6 Ignition Fuel Valve and Ignition Monitor Valve closed.

Figure 4-7

FLIGHT CONTROL SYSTEM

FROM FROM FROM
FUEL LOX FUEL
TANK TANK TANK

RETURN
TO
GROUND

GROUND
CHECKOUT
VALVE

SERVOACTUATOR
RETURN DUCT

ENGINE
OUTRIGGER

SERVOACTUATOR
SUPPLY DUCT

CHECK
VALVE

FLIGHT
SUPPLY
DUCT

FILTER MANIFOLD

FILTER

GROUND
SUPPLY
DUCT

CHECK
VALVE

⬤⬤⬤ RJ-1
▨▨▨ RP-1
▧▧▧ LOX
.... RETURN FLOW

Figure 4-8

PROPELLANTS

Propellants for the S-IC stage are RP-1 (fuel) and liquid oxygen (lox). The propellant system includes hardware for fuel fill and drain operations, tank pressurization prior to and during flight, and delivery of propellants to the engines. The system is divided into two systems, the fuel system and the lox system. Figures 4-11 through 4-13 illustrate the system operation.

FUEL LOADING AND DELIVERY

Fuel loading starts approximately thirty days before launch (figure 4-12) and continues at a rate of 2000 gpm until 99% full and then uses a 200 gpm rate until the total mass load reaches 102% of the desired load. At T-60 minutes the propellant management GSE gives the command to begin fuel level adjustment to the prescribed flight load level. This initiates a limited drain. The fuel loading probe (figure 4-11),

senses the mass level. The fuel vent and relief valve is opened during gravity drain but must be closed for pressurized drain.

RP-1 Pressurization

Fuel tank pressurization (figure 4-11) is required from engine starting through stage flight to establish and maintain a net positive suction head at the fuel inlet to the engine turbopumps. Ground supplied helium for prepressurization is introduced into the cold helium line downstream from the flow controller resulting in helium flow through the engine heat exchanger and the hot helium line to the fuel tank distributor. During flight, the source of fuel tank pressurization is helium from storage bottles mounted inside the lox tank.

Fuel tank pressure is maintained by an over-pressure switch which controls the fuel vent and relief valve, a prepressurization switch which controls GSE pressure supply

HYDRAULIC SERVOACTUATOR

ENGINE ATTACH END

SERVOVALVE

EXTEND PRESSURE

RETRACT PRESSURE

STAGE
ATTACH END

TYPICAL SERVOACTUATOR	
LENGTH	5 FEET
WEIGHT	300 POUNDS
STROKE	11 INCHES

Figure 4-9

during filling and prepressurization, and a flight prepressurization switch. During flight the fuel tank pressure is maintained by five solenoid valves located in the helium flow controller. Valve No. 1 is opened first before liftoff. During flight, valve Nos. 2, 3 and 4 are sequenced open by the I.U. to maintain tank pressure as the helium bottle pressure decreases. Valve No. 5 is opened by the flight pressurization switch if the tank pressure falls below 19.5 psia. It closes at 21.5 psia.

The onboard helium storage bottles are filled through a filtered fill and drain line upstream from the flow controller. The storage bottles are filled to a pressure of 1400 psi prior to lox loading. Fill is completed to 3150 psi after lox loading when the bottles are cold. A mechanical relief valve, set to open at 4300 psig and close at 3600 psig, provides positive pressure relief. The valve is plumbed to exhaust into the fuel tank.

RP-1 Delivery

Fuel feed (figure 4-11) is accomplished through two 12-inch ducts which connect the fuel tank to each F-1 engine. The ducts are equipped with gimbaling and sliding joints to compensate for motions from engine gimbaling and stage stresses. Prevalves, one in each fuel line, serve as an emergency backup to the main engine fuel shutoff valves. The prevalves also house flowmeters which provide flowrate data via telemetry to GSE. A fuel level engine cutoff sensor, in the bottom of the fuel tank, initiates engine shutdown when fuel is depleted if the lox sensors have failed to cut off the engines.

LOX LOADING AND DELIVERY

As the oxidizer in the bi-propellant propulsion system, lox is contained and delivered through a separate tank and delivery system (figure 4-13). The 345,000 gallon tank is filled through two 6-inch fill and drain lines. Shortly after T-6 hours lox loading begins. Three fill rates are used sequentially; a 300 gpm for tank chilldown, a 1500 gpm slow fill rate to stabilize the liquid level and thus prevent structural damage, and a fast fill rate of 10,000 gpm. At approximately 95% full, the rate is reduced to 1500 gpm and ceases when the lox loading level sensor automatically stops the fill mode. Lox boiloff is replenished at a maximum rate of 500 gpm until prepressurization occurs.

Lox Drain

The lox is drained through the two fill and drain intertank lines (figure 4-13) and an aft fill and drain line in the thrust structure. The aft fill and drain line is not used until 6.5% load is reached. During lox drain, positive ullage pressure is maintained by a GSE pressure source and two vent valves which are kept closed except when overpressure occurs.

Prior to launch, boil off in the lox tank may be harmlessly vented overboard. However, excessive geysering from boiling in the lox suction ducts can cause structural damage, and high lox temperatures near the engine inlets may prevent normal engine start. The lox bubbling system (figure 4-13) eliminates geysering and maintains low pump inlet temperatures. The helium induced convection currents circulate lox through the suction ducts and back into the tank. Once established, thermal pumping is self sustaining and continues until the interconnect valves are closed just prior to launch.

Figure 4-10

Lox Pressurization System

LOX tank prepressurization (Figures 4-12 and 4-13) is required to ensure proper engine turbopump net positive suction head during start, thrust buildup, and mainstage.

LOX tank pressure is monitored by a prepressurization switch, an absolute pressure switch and a gage pressure switch. The prepressurization switch maintains the ullage pressure between 24.2 and 26.5 psia until launch commit by controlling the GSE pneumatic supply. The absolute pressure switch and gage pressure switch limit the maximum tank pressure during flight by controlling the lox vent and relief and the lox vent valve respectively. The absolute pressure switch limit in 30 psia. The gage pressure switch limit is 25 psig. The absolute pressure switch controls for about the first 80 seconds of flight. At about this time the 30 psia limit and the 25 psig limit become equal. Beyond this point the gage pressure switch controls to limit the tank differential pressure to 25 psi.

The pressurization gas used during prepressurization is helium. Gox is used for pressurizing the lox tank during flight. A portion of the lox supplied to each engine is diverted from the lox dome into the engine heat exchanger where the hot turbine exhaust transforms lox into gox. The heated gox is delivered through the gox pressurization line and flow control valve to the lox tank. A sensing line provides pressure feedback to the flow control valve to regulate the gox flow rate and maintain the ullage pressure between 18 and 20 psia.

Lox Delivery

Lox is delivered to the engines through five suction lines (figure 4-13). The ducts are equipped with gimbals and sliding joints to compensate for motions from engine gimbaling and stage stresses. Pressure volume compensating ducts ensure constant lox flowrate regardless of the gimbaled position of the engine. Each suction line has a lox prevalve which is a backup to the engine lox valve. The prevalve cavity is charged with helium and functions as an accumulator to absorb engine induced pulses (figure 4-4).

ELECTRICAL

The electrical power system of the S-IC stage is made up of two basic subsystems: the operational power subsystem and the measurements power subsystem. Onboard power is supplied by two 28-volt batteries located as shown in figure 4-14. Battery characteristics are listed in figure 4-15.

In figure 4-16, battery number 1 is identified as the operational power system battery. It supplies power to

FUEL SUBSYSTEMS

HELIUM STORAGE BOTTLES

LOX TANK

MECHANICAL RELIEF VALVE

HELIUM FLOW CONTROLLER

HELIUM BOTTLE FILL AFT UMBILICAL NO. 2

VENT & RELIEF VALVE

PRESSURE SWITCHES (3)

LOADING PROBE

FLIGHT CONTROL VALVES

CONTINUOUS LEVEL PROBE

DISTRIBUTOR

FUEL TANK

SLOSH PROBES

ENGINE CUTOFF SENSOR

6 INCH FILL AND DRAIN VALVE AFT UMBILICAL NO. 2

COLD HELIUM

FUEL PREVALVES

GSE HELIUM SUPPLY-PREPRESSURIZATION (AFT UMBILICAL NO. 2)

HOT HELIUM

MAIN FUEL VALVE (2)

HEAT EXCHANGER

HELIUM
FUEL

1 PRESSURIZED RP-1 DRAIN REQUIRES 3 HOURS MINIMUM

Figure 4-11

S-IC PROPELLANT LOAD AND OPERATIONAL SEQUENCE

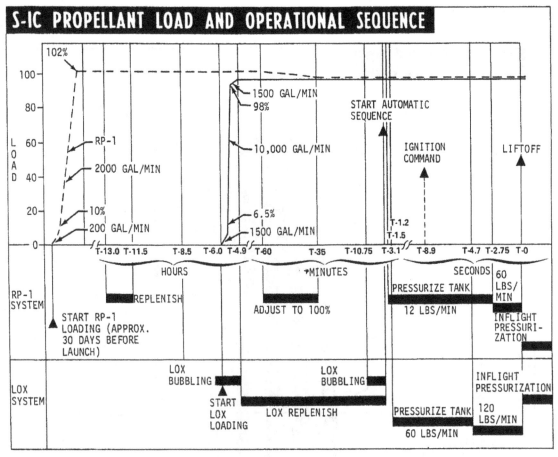

Figure 4-12

operational loads such as valve controls, purge and venting systems, pressurization systems, and sequencing and flight control. Battery number 2 is identified as the measurement power system battery. It supplies power to measurements loads such as telemetry systems, transducers, multiplexers, and transmitters. Both batteries supply power to their loads through a common main power distributor but each system is completely isolated from the other.

During the prelaunch checkout period power for all electrical loads, except range safety receivers, is supplied from GSE. The range safety receivers are hardwired to batteries 1 and 2 in order to enhance the safety and reliability of the range safety system. At T-50 seconds a ground command causes the power transfer switch to transfer the S-IC electrical loads to onboard battery power. However, power for engine ignition and for equipment heaters (turbopump and lox valves) continues to come from the GSE until terminated at umbilical disconnect.

DISTRIBUTORS

There are six power distributors on the S-IC stage. They facilitate the routing and distribution of power and also serve as junction boxes and housing for relays, diodes, switches and other electrical equipment.

There are no provisions for switching or transferring power between the operational power distribution system and the measurement power system. Because of this isolation, no failure of any kind in one system can cause equipment failure in the other system.

Main Power Distributor

The main power distributor contains a 26-pole power transfer switch, relays, and the electrical distribution busses. It serves as a common distributor for both operational and measurement power subsystems. However, each of these systems is completely independent of the other. The power load is transferred from the ground source to the flight batteries at T-50 seconds. Inflight operation of the multicontact make-before-break power transfer switch is prevented by a brake, by mechanical construction, and by electrical circuitry. Operation of the switch several times during countdown verifies performance of the brake, motor, contacts, and mechanical components.

Sequence and Control Distributor

The sequence and control distributor accepts command signals from the switch selector and through a series of

LOX SYSTEM COMPONENTS

PRESSURE SWITCHES
LOX VENT VALVE

DISTRIBUTOR
LOX VENT AND RELIEF VALVE

OXIDIZER TANK

LOX FILL AND DRAIN INTERTANK LINES

DUCT NO. 3
DUCT NO. 5
DUCT NO. 4

LOX CUTOFF SENSORS (TYPICAL 5 DUCTS)

FUEL TANK

PRESSURE SENSING LINE

HELIUM BUBBLING IN NO. 1 & 3 INDUCES CIRCULATION BY CAUSING THE LOX IN THESE DUCTS TO RISE THEREBY DRAWING LOX DOWN THROUGH THE OTHER THREE DUCTS.

SHUTOFF VALVE

PRESSURIZATION LINE

LOX BUBBLING AFT UMBILICAL NO. 3

HELIUM

LOX SUCTION LINE

DUCT NO. 1
DUCT NO. 2

PREVALVE

INTERCONNECT VALVE

PREPRESSURE (HELIUM) AFT UMBILICAL NO. 1

LOX AFT FILL AND DRAIN LINE

GOX FLOW CONTROL VALVE

MAIN LOX VALVE (2)

GOX INPUTS (TYPICAL 5 ENGINES)

HEAT EXCHANGER

F-1 ENGINE NO. 5

(TYPICAL 5 PLACES)

LEGEND
SENSE LINE
LOX
HELIUM
FUEL
GOX

Figure 4-13

S-IC ELECTRICAL POWER EQUIPMENT LOCATIONS

Figure 4-14

S-IC BATTERY CHARACTERISTICS	
TYPE	DRY CHARGE
MATERIAL	ZINC-SILVER OXIDE
ELECTROLYTE	POTASSIUM HYDROXIDE (KOH) IN PURE WATER
CELLS	20 WITH TAPS FOR SELECTING 18 OR 19 TO REDUCE OUTPUT VOLTAGE AS REQUIRED
NOMINAL VOLTAGE	1.5 VDC PER CELL: 28 + 2 VDC PER 18 TO 20 CELL GROUP
CURRENT RATINGS	BATTERY NO. 1 - OPERATIONAL LOADS = 640 AMPERE/MINUTE BATTERY NO. 2 - MEASUREMENT LOADS = 1250 AMPERE/MINUTE
GROSS WEIGHT	BATTERY NO. 1 = 22 LBS. BATTERY NO. 2 = 55 LBS.

Figure 4-15

magnetically latching relays provides a 28-volt dc command to initiate or terminate the appropriate stage function. The input from the switch selector latches a relay corresponding to the particular command. A 28-volt dc signal is routed through the closed contacts of the relay to the stage components being commanded. The relays, one for each command function, may be unlatched by a signal from GSE.

The normally closed contacts of the relays are connected in series. A 28-volt dc signal is routed through the series connected relay contacts to indicate to GSE when all sequence and control relays are in the reset state.

Propulsion Distributor

The propulsion distributor contains relays, diodes, and printed circuit boards for switching and distributing propulsion signals during launch preparation and flight.

Thrust OK Distributor

The thrust OK distributor contains relays and printed circuit assemblies which make up the thrust OK logic networks and timers required to monitor engine thrust OK pressure switches and initiate engine shutdown. Signals from two of the three thrust OK pressure switches on a particular engine will result in an output from a two-out-of-three voting network. This output activates a 0.044 second timer. If the thrust OK condition is missing longer than 0.044 seconds the timer output sends a signal to initiate engine shutdown.

Timer Distributor

Circuits to time the operation of relays, valves, and other electromechanical devices are mounted in the timer distributor.

Measuring Power Distributor

Each regulated 5-volt dc output from the seven measuring power supplies is brought to an individual bus in the measuring power distributor and then routed to the measuring and telemetry systems.

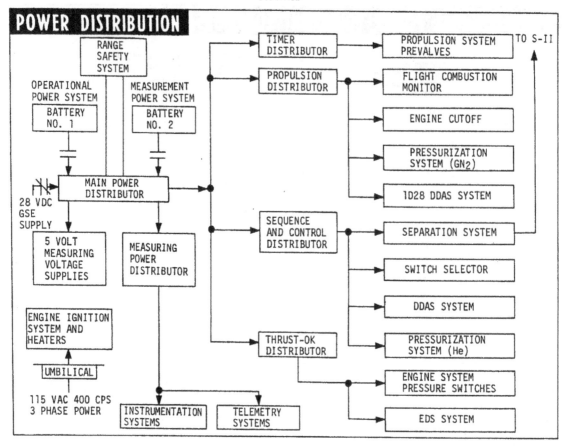

POWER DISTRIBUTION

Figure 4-16

SWITCH SELECTOR

The S-IC stage switch selector is the interface between the LVDC in the IU and the S-IC stage electrical circuits. Its function is to sequence and control various flight activities such as TM calibration, retrorocket initiation, and pressurization as shown in figure 4-17.

A switch selector is basically a series of low power transistor switches individually selected and controlled by an eight-bit binary coded signal from the LVDC in the IU. A coded word, when addressed to the S-IC switch selector, is accepted and stored in a register by means of magnetically latching relays. The coded transmission is verified by sending the complement of the stored word back to the LVDC in the IU. At the proper time an output signal is initiated via the selected switch selector channel to the appropriate stage operational circuit. The switch selector can control 112 circuits.

LVDC commands activate, enable, or switch stage electrical circuits as a function of elapsed flight time. Computer commands include:

1. Telemetry calibration.

2. Remove telemetry calibration.

3. Open helium flow control valve No. 2.

4. Open helium flow control valve No. 3.

5. Open helium flow control valve No. 4.

6. Enable center engine cutoff.

7. Enable outboard engine cutoff.

8. Arm EBW firing unit, retrorockets, and separation system.

9. Fire EBW firing unit, retrorockets, and separation system.

10. Measurement switchover.

In addition, a command from the emergency detection system in the IU can shut down all S-IC stage engines.

INSTRUMENTATION

The S-IC stage instrumentation system monitors functional operations of stage systems and provides signals for vehicle tracking during the S-IC burn. Prior to liftoff, measurements are telemetered by coaxial cable to ground support

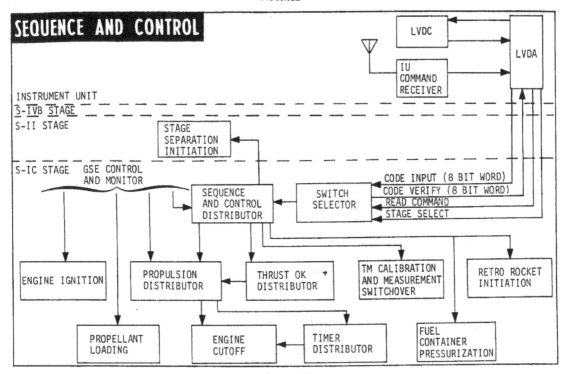

SEQUENCE AND CONTROL

Figure 4-17

equipment. During flight, data is transmitted to ground stations over RF links. Section VII provides a detailed discussion of the telemetry system.

TELEMETRY SYSTEM

The telemetry system accepts the signals produced by the measuring portion of the instrumentation system and transmits them to ground stations. The telemetry equipment includes multiplexers, subcarrier oscillators, amplifiers, modulators, transmitters, and a two-element antenna system. The telemetry system uses multiplex techniques (time sharing) to transmit large quantities of measurement data over a relatively small number of basic RF links.

There are two basic types of telemetry systems in the S-IC stage (figure 4-18). One pulse amplitude modulated/frequency modulated (PAM/FM/FM) link is used for telemetering low-to-medium frequency data such as pressure or temperature. Time multiplexed data from the PAM links are also routed through the PCM links at one third sampling rate for DDAS transmission during preflight testing and for redundant RF transmission during flight. A pulse code modulated/digital data acquisition system (PCM/DDAS) link provides for acquisition of analog and digital flight data, provides a hardwire link for obtaining PCM data and PAM time multiplexed data during test and checkout and permits the redundant monitoring of PAM data during flight. The PCM/DDAS system assembles and formats PCM/FM time shared data so it can be sent over coaxial cables for automatic ground checkout or over an RF link during flight.

The Remote Digital Submultiplexer (RDSM) provides additional data-handling capability to the PCM telemetry system. It can accept a maximum of 100 inputs which are sampled sequentially in groups of ten. The assembly handles only digital information. The inputs and outputs are set voltage levels that represent liquid level measurements and

discrete data from other sources.

MEASUREMENT SYSTEM

The measurement system senses performance parameters and feeds signals to the telemetry system. It includes transducers, signal conditioning, and distribution equipment necessary to provide the required measurement ranges and suitably scaled voltage signals to the inputs of the telemetry system.

The S-IC measuring system performs three main functions:

1. Detection of the physical phenomena to be measured and transformation of these phenomena into electrical signals.

2. Process and condition the measured signals into the proper form for telemetering.

3. Distribution of the data to the proper channel of the telemetry system.

Remote Automatic Calibration System (RACS)

The RACS is used to verify measurement circuit operation and continuity by stimulating the transducer directly, or by inserting a simulated transducer signal in the signal conditioner circuit. Measurement operation is verified at 80 percent of the maximum transducer range (high level), at 20 percent of the maximum range (low level), and at the normal run level.

ORDNANCE

The S-IC ordnance systems include the propellant dispersion (flight termination) system (figure 4-19) and the retrorocket system (figure 4-20).

TELEMETRY SYSTEMS

Figure 4-18

PROPELLANT DISPERSION

CDF TEES

POS II

CONFINED DETONATING FUSE ASSEMBLY (2 REQUIRED)

POS I

SAFETY AND ARMING DEVICE

RANGE SAFETY ANTENNA

COMMAND DESTRUCT CONTAINER

CDF ASSEMBLY (2 REQUIRED)

ORDNANCE COWLING FOR RP-1 TANK FLSC

POS III

POS IV

RANGE SAFETY ANTENNA

ORDNANCE COWLING FOR LOX TANK FLSC

CDF ASSEMBLY (2 REQUIRED)

CDF/FLSC CONNECTOR

CDF TEES

COMMAND DESTRUCT CONTAINER

LOX TANK FLSC ASSEMBLY

CONFINED DETONATING FUSE ASSEMBLIES

CDF/FLSC CONNECTOR

RP-1 TANK FLSC ASSEMBLY

HYBRID POWER DIVIDER

RANGE SAFETY COMMAND RECEIVER NO. 1

RANGE SAFETY COMMAND RECEIVER NO. 2

RANGE SAFETY DECODER

RANGE SAFETY DECODER

BATTERY NO. 1

CONTROLLER

CONTROLLER

BATTERY NO. 2

NO-SAFING PLUG

ENGINE SHUTDOWN

NO-SAFING PLUG

EBW FIRING UNIT

EBW FIRING UNIT

PROPELLANT DISPERSION BLOCK DIAGRAM

EBW DETONATOR

SAFETY AND ARMING DEVICE

CDF ASSEMBLIES

CDF TEE

CDF/FLSC CONNECTOR

RP-1 TANK FLSC ASSEMBLY

LOX TANK FLSC ASSEMBLY

Figure 4-19

4-21

Figure 4-20

PROPELLANT DISPERSION SYSTEM

The S-IC propellant dispersion system (PDS) provides the means of terminating the flight of the Saturn V if it varies beyond the prescribed limits of its flight path or if it becomes a safety hazard during the S-IC boost phase. The system is installed on the stage in compliance with Air Force Eastern Test Range (AFETR) Regulation 127.9 and AFETR Safety Manual 127.1.

The PDS is a dual channel, parallel redundant system composed of two segments. The radio frequency segment receives, decodes, and controls the propellant dispersion commands. The ordnance train segment consists of two exploding bridgewire (EBW) firing units, two EBW detonators, one safety and arming (S&A) device (shared by both channels), six confined detonating fuse (CDF) assemblies, two CDF tees, two CDF/flexible linear shaped charge (FLSC) connectors, and two FLSC assemblies.

The S&A device (figure 4-21) is a remotely controlled electro-mechanical ordnance device that is used to make safe and to arm the S-IC, S-II, and S-IVB stage PDS's. The device can complete and interrupt the explosive train by remote control, provide position indications to remote monitoring equipment, and provide a visual position indication. It also has a manual operation capability. The S&A device consists of a rotary solenoid assembly, a metal rotor shaft with two explosive inserts, and position sensing and command switches that operate from a rotor shaft cam. In the safe mode, the longitudinal axis of the explosive inserts are perpendicular to the detonating wave path, thus forming a barrier to the explosive train. To arm the device, the shaft is rotated 90 degrees to align the inserts between the EBW detonators and the CDF adapters to form the initial part of the explosive train.

Should emergency flight termination become necessary, two coded radio frequency commands are transmitted to the launch vehicle by the range safety officer. The first command arms the EBW firing units and initiates S-IC engine cutoff. (See figure 4-19 for block diagram of the PDS and location of PDS components.) The second command, which is delayed to permit charging of the EBW firing units, discharges the EBW firing units across the exploding bridgewire in the EBW detonators mounted on the S&A device (see figure 4-21). The resulting explosive wave propagates through the S&A device inserts to the CDF assemblies and to the CDF tees. The CDF tees propagate the wave through insulated CDF assemblies to the FLSC assemblies mounted on the lox and RP-1 tanks. The FLSC's provide the explosive force to longitudinally sever the propellant tanks and disperse the propellants. There are six 88-inch FLSC sections mounted on the lox tank and three 88-inch sections on the fuel tank. These sections are positioned on the propellant tanks to minimize mixing of the propellants after the tanks are severed.

RETROROCKETS

The S-IC retrorockets are mounted, in pairs, (figure 4-20) in the fairings of the F-1 engine. At retrorocket ignition the forward end of the fairing is burned and blown through by the exhausting gases. Each retrorocket is pinned securely to the vehicle support and pivot support fittings at an angle of 7.5 degrees from center line. The thrust level developed by seven retrorockets (one retrorocket out) is adequate to separate the S-IC stage a minimum of six feet from the vehicle in less than one second.

Figure 4-21

The eight retrorockets (figure 4-20), provide separation thrust after S-IC burnout. The firing command originates in the Instrument Unit and activates redundant firing systems. Additional redundancy is provided by interconnection of the two confined detonating fuse (CDF) manifolds with CDF assemblies. The exploding bridgewire (EBW) firing unit circuits are grounded by a normally closed relay until the firing command is initiated by the Instrument Unit. High voltage electrical signals are released from the two EBW firing units to the EBW detonators upon vehicle deceleration of 0.5g. The signals cause the detonator bridgewires to explode, thereby detonating the surrounding explosives. The explosion then propagates through the CDF manifold explosive and CDF assemblies into the igniter assembly. The igniter assembly located within the base of each retrorocket is then ignited, causing a buildup and release of the gases into the main grain of the retrorocket. Each retrorocket is ignited by either of two CDF initiators mounted on its aft structure. Operational ground check of the system through the firing unit is accomplished through use of pulse sensors which absorb the high voltage impulse from the firing unit and transmit a signal through the telemetry system. The pulse sensors are removed prior to launch.

Each retrorocket is a solid propellant rocket with a case bonded, twelve-point star, internal burning, composite propellant cast directly into the case and cured. The propellant is basically ammonium perchlorate oxidizer in a polysulfide fuel binder. The motor is 86 inches long by 15-1/4 inches diameter and weights 504 pounds, nominal, of which 278 pounds is propellant.

MAJOR DIFFERENCES

1▶ *On vehicles SA-508 and on, th S-IC actuators are equipped with a ±3 degree position indicator to provide position information to ESE for detection of actuator hard-over failure prior to launch commit.*

2▶ *On vehicles SA-507 and on, the ODOP system is not installed.*

This Space Intentionally Left Blank

SECTION V

S-II STAGE

TABLE OF CONTENTS

INTRODUCTION

The S-II stage provides second stage boost for the Saturn V launch vehicle. The stage (figure 5-1) is 81.5 feet long, 33 feet in diameter, and is powered by five liquid propellant J-2 rocket engines which develop a nominal vacuum thrust of 230,000 pounds each for a total of 1,150,000 pounds. The four outer J-2 engines are equally spaced on a 17.5 foot diameter circle and are capable of being gimbaled through a plus or minus 7.0 degree pattern for thrust vector control. The fifth engine is mounted on the stage centerline and is fixed. Stage and engine characteristics are listed in figure 1-3.

At engine cutoff the S-II stage separates from the S-IVB and, following a suborbital path, reenters the atmosphere where it disintegrates due to reentry loads.

The stage consists of the structural airframe, the J-2 engines, piping, valves, wiring, instrumentation, electrical and electronic equipment, ordnance devices, and four solid propellant ullage rockets. These are collected into the following major systems: structural, environmental control, propulsion, flight control, pneumatic control, propellant, electrical, instrumentation, and ordnance. The stage has structural and electrical interfaces with the S-IC and S-IVB stages; and electrical, pneumatic, and fluid interfaces with ground support equipment through its umbilicals and antennae.

STRUCTURE

The S-II airframe (figure 5-1) consists of a body shell structure (forward and aft skirts and interstage), a propellant tank structure (liquid hydrogen and liquid oxygen tanks), and a thrust structure. The body shell structure transmits first and second stage boost loads (axial, shear, and bending moment) and stage body bending and longitudinal forces between the adjacent stages, the propellant tank structure, and the thrust structure. The propellant tank structure holds the propellants, liquid hydrogen (LH$_2$) and liquid oxygen (lox), and provides structural support between the aft and forward skirts. The thrust structure transmits the thrust of the five J-2 engines to the body shell structure; compression loads from engine thrust; tension loads from idle engine weight; and cantilever loads from engine weight during S-II boost.

BODY SHELL STRUCTURE

The body shell structure units, the forward skirt, aft skirt, and interstage are of the same basic design except that the aft skirt and interstage are of generally heavier construction because of higher structural loads.

Each unit is a cylindrical shell of semimonocoque construction, built of 7075 aluminum alloy material, stiffened by external hat-section stringers and stabilized internally by circumferential ring frames. The forward skirt has a basic skin thickness of 0.040 inch and the aft skirt and interstage both have basic skin thicknesses of 0.071 inch.

THRUST STRUCTURE

The thrust structure, like the body shell structure, is of semimonocoque construction but in the form of a truncated cone increasing in size from approximately 18 feet in diameter to the 33 foot outside diameter of the airframe. It is stiffened by circumferential ring frames and hat-section stringers. Four pairs of thrust longerons (two at each outboard engine location) and a center engine support beam cruciform assembly accept and distribute the thrust loads of the J-2 engines. The shell structure is of 7075 aluminum alloy. A fiberglass honeycomb heat shield, supported from the lower portion of the thrust structure, protects the stage base area from excessive temperatures during S-II boost.

The conical shell also serves to support the major portion of systems components carried on the S-II, either mounted in environmentally controlled equipment containers or directly to the airframe structure (See figure 5-2).

PROPELLANT TANK STRUCTURE

The LH$_2$ tank consists of a long cylinder with a concave modified ellipsoidal bulkhead forward and a convex modified ellipsoidal bulkhead aft. The aft bulkhead is common to the lox tank. The LH$_2$ tank wall is composed of six cylindrical sections which incorporate stiffening members in both the longitudinal and circumferential directions. Wall sections and bulkheads are all fabricated from 2014 aluminum alloy and are joined together by fusion welding. The forward bulkhead has a 36 inch diameter access manhole built into its center.

The common bulkhead is an adhesive-bonded sandwich assembly employing facing sheets of 2014 aluminum alloy and fiberglass/phenolic honeycomb core to prevent heat transfer and retain the cryogenic properties of the two fluids to which it is exposed. Fiberglass core insulation thickness varies from approximately 5 inches at the apex to 0.080 inch at the outer extremity. No connections or lines pass through the common bulkhead. The forward skin has a "J" section return at the outer edge to permit peripheral attachment to the LH$_2$ tank while the lower facing is carried through to provide structural continuity with the lox tank aft bulkhead.

The liquid oxygen tank (figure 5-3) consists of ellipsoidal fore and aft halves with waffle-stiffened gore segments. The tank is fitted with three ring-type slosh baffles to control

This Space Intentionally Left Blank

S-II STAGE STRUCTURE

SYSTEMS TUNNEL

VEHICLE STATION 2519

VEHICLE STATION 1541

81-1/2 FEET

33 FEET

FORWARD SKIRT

11-1/2 FEET

LIQUID HYDROGEN TANK (37,737 CU FT)

56 FEET

LH₂/LOX COMMON BULKHEAD

LIQUID OXYGEN TANK (12,745.5 CU FT)

22 FEET

AFT SKIRT THRUST STRUCTURE

14-1/2 FEET

INTERSTAGE

18-1/4 FEET

Figure 5-1

5-2

This Space Intentionally Left Blank

S-II STAGE EQUIPMENT LOCATIONS -TYPICAL

1. RANGE SAFETY COMMAND CONTAINER
2. THERMAL CONTROL MANIFOLD
3. LIQUID HYDROGEN TANK BULKHEAD
4. INSTRUMENTATION CONTAINER
5. INSTRUMENTATION CONTAINER
6. FORWARD SKIRT
7. MANHOLE COVER
8. RADIO COMMAND ANTENNA - 4 PLACES
9. TELEMETRY ANTENNA - 4 PLACES
10. UMBILICAL AND SERVICE CONNECTIONS
11. THRUST STRUCTURE
12. INSTRUMENTATION CONTAINER
13. INSTRUMENTATION CONTAINER
14. INSTRUMENTATION CONTAINER
15. PROPELLANT MANAGEMENT PACKAGE
16. ELECTRICAL ASSY (TIMING AND STAGING CENTER)
17. SERVOACTUATOR (TYPICAL)
18. ULLAGE ROCKET - 4 PLACES
19. J-2 ENGINE (NO. 2)
20. HEATSHIELD
21. ACCESS DOOR
22. ELECTRICAL POWER AND CONTROL SYSTEM
23. FLIGHT CONTROL CONTAINER
24. INTERSTAGE
25. LOX FEEDLINE
26. SIGNAL CONDITIONER CONTAINER
27. AFT SKIRT
28. INSTRUMENTATION CONTAINER
29. SIGNAL CONDITIONER CONTAINER
30. INSTRUMENTATION CONTAINER
31. SYSTEMS TUNNEL
32. PROPELLANT DISPERSION SYSTEM
33. LH2 VENTLINE AND VALVES
34. PROPELLANT MANAGEMENT ELECTRICAL CONTAINER
35. TELEMETRY EQUIPMENT CONTAINER
36. TELEMETRY EQUIPMENT CONTAINER
37. SIGNAL CONDITIONER CONTAINER

Figure 5-2

propellant sloshing and minimize surface disturbances and cruciform baffles to prevent the generation of vortices at the tank outlet ducts and to minimize residuals. A six-port sump assembly located at the lowest point of the lox tank provides a fill and drain opening and openings for five engine feed lines.

SYSTEMS TUNNEL

A systems tunnel, housing electrical cables, pressurization lines, and the tank propellant dispersion ordnance, is attached externally from the S-II stage aft skirt area to the forward skirt. It has a semicircular shape 22 inches wide and about 60 feet long. Cabling which connects the S-IC stage to the instrument unit also runs through this tunnel (see figures 5-1 and 5-2).

ENVIRONMENTAL CONTROL

The environmental control system is supplied dehumidified, thermally-conditioned air and nitrogen from a ground source for temperature control and purging of the compartments during prelaunch operations only.

THERMAL CONTROL

The thermal control system (figure 5-2) provides temperature control to forward and aft skirt mounted equipment containers. The system is put into operation shortly after the vehicle is mated to pad facilities. Air is used as the conditioning medium until approximately two hours prior to cryogenic loading. At this time gaseous nitrogen (GN$_2$) is used until umbilical disconnect to preclude the possibility of an explosion in the event of LH$_2$ leakage. The change to GN$_2$ is made before propellant loading to ensure that all oxygen is expelled and dissipated before a hazard can arise. The nitrogen flow is terminated at liftoff, and no flow is provided during boost, since the equipment container insulation is capable of maintaining equipment temperatures throughout the S-II flight trajectory.

ENGINE COMPARTMENT CONDITIONING

The engine compartment conditioning system purges the engine and interstage areas of explosive mixtures and maintains a proper temperature. Purging the compartment is accomplished prior to propellant tanking and whenever

Figure 5-3

propellants are on board. A 98% GN_2 atmosphere circulating through the compartment maintains desired temperature while the danger of fire or explosion resulting from propellant leakage is minimized.

TANK INSULATION

All external surfaces of the LH_2 tank require insulation to prevent condensation and to reduce temperature rise during cryogenic operations. The insulation material varies from area to area on the stage. A foam-filled honeycomb material (figure 5-4) is used as the core of the LH_2/lox tank common bulkhead. This honeycomb insulation has a network of passages through which helium gas is forced for purging and leak detection. The bolting ring, LH_2 tank sidewalls and forward bulkhead, and the forward skirt are all covered with a spray-on polyurethane foam. In addition, quarter-inch thick cork sheet is bonded over the polyurethane foam in certain areas. These are the multiple protuberance wake effect areas around the LH_2 feedline fairings, recirculation return line fairings, LH_2 fill and drain fairings, and systems tunnel. Also a protective cork strip is installed on the ramp of the forward skirt foam insulation.

PURGE AND LEAK DETECTION

From initiation of propellant loading, right up to umbilical disconnect, the purge and leak detection system provides a flow of helium through the honeycomb core of the LH_2/lox tank common bulkhead and through the adjacent J-ring area. This purge is for the purpose of excluding hazardous gases. A hazardous gas analyzer monitors the helium outflow as shown in figure 5-5. If hazardous quantities of hydrogen/oxygen are present an alarm is given through the GSE.

PROPULSION

The S-II stage engine system consists of five single-start J-2 rocket engines utilizing liquid oxygen and liquid hydrogen for propellants. The four outer J-2 engines are mounted

parallel to the stage centerline. These engines are suspended by gimbal bearings to allow thrust vector control. The fifth engine is fixed and is mounted on the centerline of the stage.

HONEYCOMB INSULATION

TEDLAR SEAL

POLYURETHANE FOAM

ADHESIVE

NYLON-PHENOLIC LAMINATED SKIN

HONEYCOMB CORE

PURGE GROOVES (PASSAGES)

ADHESIVE

TANK SKIN

Figure 5-4

INSULATION PURGE AND LEAK DETECTION

GHe

LH2 TANK

GSE PURGE CONTROL CONSOLE

GHe

LOX TANK

GSE GAS ANALYZER

GSE VACUUM PUMP

COMMON BULKHEAD HONEYCOMB

SPRAY-ON POLYURETHANE FOAM

SHEET CORK OVER RAMP AREAS, FAIRINGS AND SYSTEMS TUNNEL

PURGED UNINSULATED AREA

Figure 5-5

J-2 ROCKET ENGINE

The J-2 rocket engine (figure 5-6) is a high performance, high altitude, engine utilizing liquid oxygen and liquid hydrogen as propellants. The only substances used in the engine are the propellants and helium gas. The extremely low operating temperature of the engine prohibits the use of lubricants or other fluids. The engine features a single tubular-walled, bell-shaped thrust chamber and two independently driven, direct drive, turbopumps for liquid oxygen and liquid hydrogen. Both turbopumps are powered in series by a single gas generator, which utilizes the same propellants as the thrust chamber. The main hydraulic pump is driven by the oxidizer turbopump turbine. The ratio of fuel to oxidizer is controlled by bypassing liquid oxygen from the discharge side of the oxidizer turbopump to the inlet side through a servovalve.

The engine valves are controlled by a pneumatic system powered by gaseous helium which is stored in a sphere inside the start tank. An electrical control system, which uses solid state logic elements, is used to sequence the start and shutdown operations of the engine. Electrical power is stage supplied.

During the burn period, the lox tank is pressurized by flowing lox through the heat exchanger in the oxidizer turbine exhaust duct. The heat exchanger heats the lox causing it to expand. The LH_2 tank is pressurized during the burn period by GH_2 from the thrust chamber fuel manifold.

Thrust vector control is achieved by gimbaling each of the outboard engines with hydraulic actuators powered by engine mounted hydraulic pumps.

ENGINE OPERATION

Engine operation includes starting, mainstage and cutoff. The starting and cutoff phases are periods of transition in which a sequence of activities occurs. These phases are developed in detail in the following paragraphs.

Start Preparations

Preparations for an engine start include ascertaining the positions and status of various engine and stage systems and components. The J-2 engine electrical control system controls engine operation by means of electrical signals. The heart of the engine electrical control system is the electrical

J-2 ROCKET ENGINE

POS II   POS IV

POS I

VIEW ROTATED 180°

LEGEND

1. GIMBAL	7. EXHAUST MANIFOLD	13. START TANK DISCHARGE VALVE	19. ANTI-FLOOD CHECK VALVE
2. FUEL INLET DUCT	8. THRUST CHAMBER	14. FUEL TURBOPUMP	20. HEAT EXCHANGER
3. OXIDIZER INLET DUCT	9. OXIDIZER TURBINE BYPASS VALVE	15. FUEL BLEED VALVE	21. PROPELLANT UTILIZATION VALVE
4. OXIDIZER TURBOPUMP	10. TURBINE BYPASS DUCT	16. GAS GENERATOR	22. PNEUMATIC CONTROL PACKAGE
5. START TANK	11. MAIN FUEL VALVE	17. ELECTRICAL CONTROL PACKAGE	
6. AUXILIARY FLIGHT INSTRUMENTATION PACKAGE	12. HIGH PRESSURE FUEL DUCT	18. PRIMARY FLIGHT INSTR. PACKAGE	

Figure 5-6

control package (17, figure 5-6) which sequences and times the engine start or cutoff functions.

Engine cutoff automatically causes the electrical control package circuitry to reset itself ready for start, providing all reset conditions are met. The LVDC issues an engine ready bypass signal just prior to an engine start attempt. This bypass signal acts in the same manner as a cutoff would act. The reset signals engine ready and this allows the LVDC to send its start command. Receipt of the start command initiates the engine start sequence.

Engine Start Sequence

When engine start is initiated (3, figure 5-7) the spark exciters in the electrical control package provide energy for the gas generator (GG) and augmented spark igniter (ASI) spark plugs (4). The helium control and ignition phase control valves, in the pneumatic control package (1), are simultaneously energized allowing helium from the helium tank (2) to flow through the pneumatic regulator to the pneumatic control system. The helium is routed through the internal check valve in the pneumatic control package (1) to ensure continued pressure to the engine valves in the event of helium supply failure. The regulated helium fills a pneumatic accumulator, closes the propellant bleed valves, (5) and purges (6) the oxidizer dome and gas generator oxidizer injector manifold. The oxidizer turbopump (12) intermediate seal cavity is continuously purged. The mainstage control valve holds the main oxidizer valve closed and opens the purge control valve which allows the oxidizer dome and gas generator oxidizer injector to be purged (6). The mainstage control valve also supplies opening control pressure to the oxidizer turbine bypass valve (13). An ignition phase control valve, when actuated, opens the main fuel valve (7) and the ASI oxidizer valve (8) and supplies pressure to the sequence valve located within the main oxidizer valve (14). Fuel is tapped from downstream of the main fuel valve for use in the ASI (4). Both propellants, under tank pressure, flow through the stationary turbopumps (12).

The sequence valve, in the main fuel valve (12), opens when the fuel valve reaches approximately 90% open and routes helium to the start tank discharge valve (STDV) (11) control valve. Simultaneously with engine start, the STDV delay timer is energized. Upon expiration of the STDV timer, and the receipt of a stage supplied mainstage enable signal, the STDV control valve and ignition phase timer are energized. As the STDV control valve energizes, the discharge valve opens, allowing pressurized GH_2 to flow through the series turbine drive system. This accelerates both turbopumps (12) to the proper operating levels to allow subsequent ignition and power build up of the gas generator (16). The relationship of fuel to lox turbopump speed buildup is controlled by an orifice in the oxidizer turbine bypass valve (13). During the start sequence the normally open oxidizer bypass valve (13) permits a percentage of the gas to bypass the oxidizer turbine.

During this period, ASI combustion occurs. Absence of a start tank depressurized signal will cause cutoff at the expiration of the ignition phase timer. Presence of the start tank depressurized signal allows the mainstage control valve to energize. Simultaneously, the sparks deenergize timer is energized and the STDV control valve is deenergized, causing the STDV to close. Helium pressure is vented from the main oxidizer valve (14) and from the purge control valve through

the mainstage control valve. The purge control valve closes, terminating the oxidizer injector manifold purges (6). Pressure from the mainstage control valve is routed to open the main oxidizer valve (14).

A sequence valve operated by the main oxidizer valve (14) permits GHe to open the gas generator control valve (4) and to close the oxidizer turbine bypass valve (13). Flow to close the oxidizer turbine bypass valve (13) is restricted as it passes through an orifice. The orifice controls the closing speed of this valve.

Propellants flowing into the gas generator, (16) are ignited by the sparkplugs (4). Combustion of the propellants cause the hot gases to drive the turbopumps (12). The turbopumps rotation causes propellant pressure to build up. The propellant flow increases and is ignited in the thrust chamber by the torch from the ASI.

Transition into mainstage occurs as the turbopumps (12) accelerate to steadystate speeds. As oxidizer pump output pressure increases a thrust OK signal is generated by either of the two thrust OK pressure switches (17). (Cutoff occurs if no., signal is received before expiration of the sparks deenergized timer.) The ASI and GG sparks exciters are deenergized at expiration of the sparks deenergized timer. Cutoff occurs if both pressure switch actuated signals (thrust OK) are lost during mainstage operation.

Steadystate operation is maintained until a cutoff signal is initiated. During this period, GH_2 is tapped from the fuel injection manifold to pressurize the LH_2 tank. The lox tank is pressurized by lox heated by the heat exchanger in the turbine exhaust duct.

Propellant utilization is discussed under a subsequent paragraph heading of PROPELLANTS. Nominal engine thrust and specific impulse as a function of mixture ratio for the engines are shown in figure 5-8.

Engine Cutoff

The S-II J-2 engine may receive cutoff signals from several different sources. These sources include engine interlock deviations, EDS automatic and manual abort cutoffs and propellant depletion cutoff. Each of these sources signal the LVDC in the IU. The LVDC sends the engine cutoff signal to the S-II switch selector. The switch selector, in turn, signals the electrical control package. The electrical control package controls all the local signals for the cutoff sequence.

Center engine cutoff (CECO) will occur approximately 70 seconds ahead of outboard engines cutoff. This early shutdown is programmed to preclude longitudinal oscillations which have occurred late in the S-II burn on previous missions.

Cutoff Sequence

The engine cutoff sequence is shown graphically in figure 5-9. The electrical control package receives the cutoff signal (1), and deenergizes the mainstage and ignition phase control valves in the pneumatic control package (2), while energizing the helium control deenergize timer. The mainstage control valve closes the main oxidizer valve (3) and opens the purge control valve and the oxidizer turbine bypass valve (5). The purge control valve directs a helium purge (11) to the oxidizer dome and GG oxidizer injector. The ignition phase control valve closes the ASI oxidizer valve (4) and the main fuel valve (5) while opening the fast shutdown valve. The fast shutdown valve now rapidly vents the return flow from the GG control valve. All valves except the ASI oxidizer valve (4)

S-II ENGINE START

FROM LOX TANK

START COMMAND FROM I U

ELECTRICAL CONTROL PACKAGE

STAGE SUPPLIED MAIN STAGE ENABLE SIGNAL

TO LH₂ TANK

FROM LH₂ TANK

TO LOX TANK

TO LH₂ TANK

FROM GSE

LEGEND

SENSE LINE
HELIUM
FUEL
COMBUSTION GASES
LOX
GASEOUS HYDROGEN

Figure 5-7 (Sheet 1 of 3)

S II ENGINE START

SEQUENCE	EVENT	TIME IN SECONDS

T3 + 1.0   3.0   5.0   7.0   9.0   11.0   13.0

PRE-IGNITION

NOTE

ALL PNEUMATIC VALVES ARE CONTROLLED BY THE PNEUMATIC CONTROL PACKAGE ⟨1⟩ WHICH IS LOCATED ON THE ENGINE. HELIUM FOR THE PNEUMATIC SYSTEM IS SUPPLIED BY THE CONTROL SPHERE INSIDE THE GH_2 START TANK ⟨2⟩.

⟨3⟩ Start Command (Occurs at T_3 + 1.4) Engine Ready → Engine Start Sequence

⟨4⟩ Augmented Spark Igniter (ASI) spark plugs and Gas Generator (GG) spark plugs fire.

⟨5⟩ Bleed valves stop return flow to propellant tanks.

⟨6⟩ Oxidizer dome and gas generator oxidizer injector are purged.

⟨7⟩ Main fuel valve allows LH_2 to flow into engine thrust chamber and into ASI.

⟨8⟩ ASI oxidizer valve allows lox to flow to ASI.

⟨9⟩ Sparks ignite the propellants in the ASI.

⟨10⟩ Stage supplied mainstage enable signal.

Figure 5-7 (Sheet 2 of 3)

Figure 5-7 (Sheet 3 of 3)

5-10

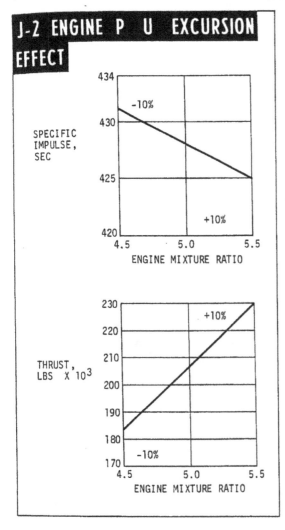

J-2 ENGINE P U EXCURSION EFFECT

Figure 5-8

and oxidizer turbine bypass valve (8), are spring loaded closed. This causes the valves to start moving closed as soon as the pressure to open them is released. GG combustion aids closing of the GG control valve.

Expiration of the helium control deenergize timer causes the helium control valve to close. When the helium control valve closes, it causes the oxidizer dome and GG oxidizer injector purges (11) to stop. An orifice in the locked up lines bleeds off pressure from the propellant bleed valves (13). This loss of pressure allows springs to open the propellant bleed valves, allowing propellants to flow back to the propellant tanks.

MALFUNCTION DETECTION

Each engine is provided with a system to detect malfunctions and to effect a safe shutdown. If neither mainstage OK pressure switch has indicated sufficient thrust for mainstage operation of the ignition phase timer, a shutdown of the particular engine is initiated. Once an engine attains mainstage operation, it is shut down if both mainstage OK pressure switches deactuate due to low level thrust.

FLIGHT CONTROL

The center engine is fixed in place while the four outer engines are gimbaled in accordance with electrical signals from the flight control computer in the IU for thrust vector control. Each outboard engine is equipped with a separate, independent, closed-loop, hydraulic control system (figure 5-10). The system includes two servoactuators mounted perpendicular to each other that provide control over the vehicle pitch, roll and yaw axes. The servoactuators are capable of deflecting the engine ± 7 degrees in the pitch and yaw planes, at the rate of 8 degrees per second.

The primary components of the hydraulic control system are an auxiliary pump, a main pump, an accumulator/reservoir manifold assembly, and two servoactuators (figures 5-10 and 5-11). The auxiliary pump is used prior to launch to maintain the hydraulic fluid temperature between 65 and 105 degrees F. The pump delivers two gallons per minute at 3650 psig, and is driven by a 400-cycle motor on GSE power.

The main pump is mounted to, and driven by, the engine lox turbopump. It is used during stage powered flight and delivers hydraulic fluid at 8 gallons per minute at 3500 psig. Prior to launch, the accumulator is pressurized with GN_2 and filled with hydraulic fluid from the pressurized auxiliary pump flow. The reservoir is, in turn, pressurized by the accumulator through a piston-type linkage. The accumulator/reservoir manifold assembly consists of a high pressure (3500 psig) accumulator which receives high pressure fluid from the pumps and a low pressure (88 psig) reservoir which receives return fluid from the servoactuators. During engine firing, hydraulic fluid is routed under pressure from the main pump to the pressure manifold of the accumulator/reservoir.

Hydraulic fluid, under pressure in the accumulator, furnishes high pressure fluid for sudden demands and smooths out pump pulsations. This pressurized hydraulic fluid is directed to the two identical, electrically controlled, hydraulically powered, servoactuators. The servoactuators have a nominal operating pressure of 3500 psig and provide the necessary forces and support to accurately position the engine in response to flight control system signals. The servoactuator is a power control unit that converts electrical command signals and hydraulic power into mechanical outputs that gimbal the engine. The developed force, direction, and velocity are determined by an electro-hydraulic servovalve.

Command signals received from the guidance system are interpreted as linear variations from a known piston position. Hydraulic fluid is then directed to either side of the actuator piston as required to satisfy the guidance command.

Actuator return fluid is routed to the reservoir which stores hydraulic fluid at sufficient pressure to supply a positive pressure at the main pump inlet.

PREFLIGHT OPERATION

During and following propellant loading, the hydraulic system fluid is intermittently recirculated by the electrically driven auxiliary pump in order to prevent the fluid from freezing. Recirculation is terminated just prior to S-IC ignition command. Recirculation is not necessary during S-IC burn, due to the short duration of the burn.

S-II ENGINE CUTOFF

ELECTRICAL CONTROL PACKAGE

CUTOFF SIGNAL FROM LVDC

FROM LOX TANK

FROM LH2 TANK

TO LH2 TANK

TO LOX TANK

TO CONTROL VALVES

PNEUMATIC CONTROL PACKAGE

TO LH2 TANK

FROM GSE

LEGEND

SENSE LINE
HELIUM
FUEL
LOX
ELECTRICAL

Figure 5-9 (Sheet 1 of 2)

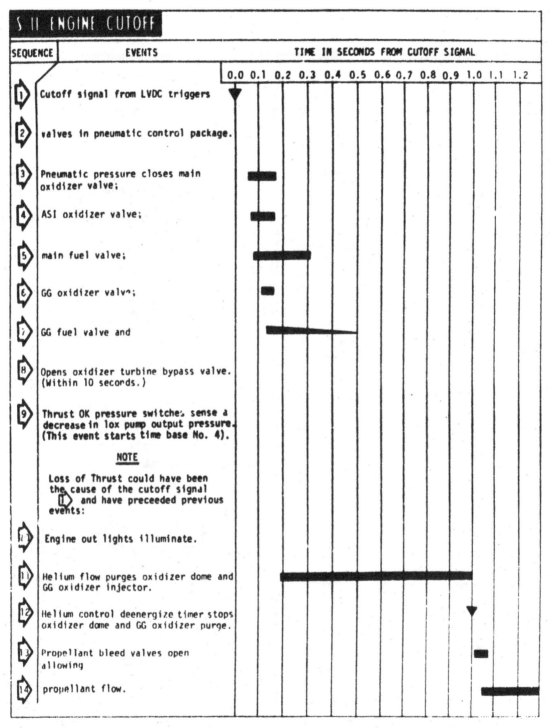

Figure 5-9 (Sheet 2 of 2)

Figure 5-10

S-II ENGINE ACTUATION

Figure 5-11

INFLIGHT OPERATION

After S-IC/S-II stage separation, an S-II switch selector command unlocks the accumulator lockup valves, releasing high pressure fluid to each of the two servoactuators. The accumulator stored fluid provides gimbaling power prior to main hydraulic pump operation. During S-II mainstage operation the main hydraulic pump supplies high pressure fluid to the servoactuators for gimbaling.

PNEUMATIC CONTROLS

The pneumatic control system (figure 5-12) consists of the ground pneumatic control system and the onboard pneumatic control system. The ground system utilizes helium supplied directly from a ground source, and the onboard system utilizes helium from onboard storage spheres.

GROUND PNEUMATICS

Ground supplied helium controls and actuates various valves during preflight operations. These include the vent valves, fill and drain valves, recirculation return line valves, and main propellant line prevalves.

ONBOARD PNEUMATICS

The onboard pneumatic control systems consist of a stage propellant valve control system and an engine pneumatic control system. The stage onboard pneumatic control system

is supplied from the helium receiver. It is pressurized to 3000 psig at approximately T-34 minutes. Pneumatic pressure from the helium receiver is regulated to 750 psig by the control regulator and is used during flight to actuate the prevalves and recirculation valves. The engine pneumatic control system is discussed in the paragraph on J-2 ROCKET ENGINE.

PROPELLANTS

The propellant systems supply fuel and oxidizer to the five J-2 rocket engines. This is accomplished by the propellant management components and the servicing, conditioning, and delivery subsystems.

PROPELLANT SERVICING SYSTEM

Pad servicing operations include the filling, draining, and purging of propellant tanks and lines as required during launch preparations.

Ground interface is through the umbilicals, to the fill and drain valves, and into the propellant tanks. Propellants then enter the engine feed lines, stopping at the closed main valves. Refer to figure 5-13 for propellant loading data. The tanks are vented by opening the tank vent valves, two per propellant tank, to allow ullage gas to escape from the tanks. Actuation pressure for the propellant tanks vent valves is provided by two separate 750-psig ground-supplied helium systems. One system actuates the lox tank vent valves, and

PNEUMATIC CONTROL SYSTEM

Figure 5-12

TYPICAL PROPELLANT LOADING DATA				
PROPELLANT	TYPE FILL	APPROXIMATE FILL RATE (GALLONS/MINUTE)	LEVEL (TANK % LEVEL)	APPROXIMATE COUNT TIME (HR:MIN:SEC)
LOX	PRECOOL FAST SLOW REPLENISH	500 5,000 1,000 0 to 100	0 to 5 5 to 98 98 to 100	T-7:04:00 to 6:54:00 T-6:54:00 to 6:30:00 T-6:30:00 to 6:27:00 T-5:45:00 to 0:03:06
LH2	PRECOOL FAST SLOW REPLENISH	1,000 10,000 1,000 0 to 480	0 to 5 5 to 98 98 to 100	T-4:54:00 to 4:36:00 T-4:36:00 to 4:16:00 T-4:16:00 to 4:11:00 T-3:33:00 to 0:03:06

Figure 5-13

the other system actuates the LH_2 tank vent valves. The vent valves are open during propellant loading operations and closed for tank pressurization.

If the launch is aborted, draining of the propellant tanks can be accomplished by pressurizing the tanks, opening the fill valves, and reversing the fill operation.

RECIRCULATION SYSTEM

Propellant recirculation is accomplished in order to maintain uniform cryogenic density and temperature and to preclude the formation of gas in propellant plumbing. This gas could cause cavitation in the turbopumps during J-2 engine start, or result in a slow engine start, slow thrust buildup or power surges.

LOX Recirculation

Lox conditioning by natural convection (figure 5-14) is initiated shortly after start of lox fill and continues until approximately T-33 minutes. At that time, helium is injected into the lox recirculation return line to boost recirculation. Helium boost is continuous until just prior to S-II ignition. After launch helium is supplied from a pressurized sphere. During recirculation, lox prevalves and recirculation return valves remain open. Return line valves are closed at termination of recirculation.

Lox conditioning is accomplished by recirculating the lox down the engine feed ducts through the prevalves, the lox turbopump, into the return lines, through the engine bleed valves, and back into the lox tank.

LH_2 Recirculation

LH_2 recirculation (figure 5-15) is initiated at approximately T-30 minutes and is terminated just prior to S-II ignition. Forced recirculation during launch and S-IC boost consists of closing the LH_2 feed line prevalves and starting the LH_2 recirculation pumps. A separate recirculation pump is provided for each of the five feed ducts.

LH_2 conditioning is accomplished by pumping the fuel through the recirculation bypass valve, into the LH_2 feed ducts downstream of the prevalves, through the LH_2 turbopump, through the LH_2 bleed valve, the recirculation return valve, and back into the fuel tank.

Recirculation is terminated by opening the prevalves, stopping the pumps, and closing the recirculation return valves.

PREPRESSURIZATION

After loading has been completed, and shortly before liftoff, the vent valves are closed and the propellant tanks are pressurized to their required levels by helium from ground supplies. Pressurization of the propellant tanks is required prior to liftoff to provide the required net positive suction head (NPSH) at the turbopump inlets for engine start. It is accomplished from a ground regulated helium source. Pressurization is initiated by the terminal countdown sequencer at approximately T-3 minutes and 7 seconds for the lox tank and T-1 minute and 37 seconds for the LH_2 tank. Pressurization is terminated at T-30 seconds for the lox tank and the LH_2 tank.

Both propellant tanks are pressurized in the same manner by separate systems (figures 5-14 and 5-15). At initiation of prepressurization, the tank vent valves are closed and the disconnect valve and ground prepressurization valves are opened to allow GHe at cryogenic temperatures to flow from the ground source through the prepressurization solenoid valve into the tank pressurization line. This line carries helium into the propellant tank through the tank gas distributor.

The vent valves act as relief valves allowing ullage gas to be vented directly overboard during flight. The lox tank vent valves limit ullage pressure to 42 psia in the lox tank. The LH_2 tank has dual setting vent valves. During the S-IC burn, LH_2 tank ullage pressure is limited to 29.5 psig and during the S-II burn to 33.0 psig.

Each propellant tank has a fill overpressure switch for personnel safety. The switch sends a signal to the GSE and is used only during loading.

PROPELLANT DELIVERY SUBSYSTEMS

The function of the engine feed systems is to transfer the liquid propellants from their tanks to the J-2 rocket engines. Each propellant tank is provided with five prevalves which provide open/close control of the flow of propellants through separate feedlines to each engine.

The prevalves are normally open, pneumatically actuated, electrically controlled, butterfly-gate type valves. Built-in four-way pneumatic control solenoids permit 750 ± 50 psig helium pressure to actuate the prevalves. Should a loss of pneumatic or electrical power occur, the prevalves are spring actuated to the open position. The prevalves remain open during S-II powered flight unless a signal is received from the engine shutdown system.

LOX FEED SYSTEM

The lox feed system furnishes lox to the five engines. This system includes four 8-inch, vacuum-jacketed feed ducts, one uninsulated feed duct, and five normally open prevalves. At engine start, lox flows from the tank, through the prevalves and feed lines, to each engine. Approximately 300 milliseconds after main valve closure, the lox prevalves are closed, providing a redundant shutoff for the lox feed system.

LH_2 FEED SYSTEM

The LH_2 feed system furnishes LH_2 to the five engines. This system includes five 8-inch vacuum-jacketed feed ducts and five normally open prevalves. The prevalves are closed following tank loading and remain closed until just prior to S-II ignition command. At engine start, LH_2 flows from the tank, through the prevalves and feed lines, to each engine. Approximately 425 milliseconds after main valve closure, the prevalves are closed, providing a redundant shutoff for the LH_2 feed system.

LOX TANK PRESSURIZATION

Lox tank pressurization (figure 5-14) is initiated at S-II ignition and continues until engine cutoff. Pressurization is accomplished with gaseous oxygen obtained by heating lox bled from the lox turbopump outlet.

LOX SYSTEM PRESSURIZATION FLOW AND CONDITIONING

Figure 5-14

LH₂ SYSTEM PRESSURIZATION FLOW AND CONDITIONING

Figure 5-15

When the turbine discharge pressure reaches a pressure differential of 100 psi, a portion of the lox supplied to the engine is diverted into the heat exchanger where it is turned into gox. The gox flows from each heat exchanger into a common pressurization duct through the tank pressurization regulator, and into the tank through the gas distributor. Approximately three minutes after S-II ignition the pressurization regulator is commanded to a full open position where it remains for the rest of the S-II burn. Ullage pressure is allowed to rise to a maximum of approximately 42 psia where the tank relief valves open preventing a further rise. Maximum pressure is reached approximately six minutes after S-II ignition. Step pressurization of the lox tank increases the NPSH to reduce pump induced engine thrust oscillations.

LH$_2$ TANK PRESSURIZATION

During S-II powered flight gaseous hydrogen (GH$_2$) for LH$_2$ tank pressurization (figure 5-15) is bled from the thrust chamber hydrogen injector manifold of each of the four outboard engines. After S-II engine ignition, liquid hydrogen is preheated in the regenerative cooling tubes of the engine and tapped off from the thrust chamber injector manifold in the form of GH$_2$ to serve as a pressurizing medium.

The GH$_2$ passes from each injector manifold into a stage manifold, through the pressurization line and tank pressurization regulator, and into the tank through the LH$_2$ tank gas distributor. The flowrate is varied according to the LH$_2$ tank ullage pressure, which is sensed by the reference pressure line connecting the LH$_2$ tank and the tank pressurization regulator.

At approximately 5 minutes after S-II engine ignition a step pressurization command from the stage switch selector activates the regulator to a fully open position, where it remains the rest of S-II boost. When the regulator is in the full open position, LH$_2$ tank pressure increases to a nominal 33 psia. Pressure in excess of 33 psia is prevented by the LH$_2$ tank vent valves. This step pressurization compensates for the loss of head pressure caused by the lowering of the fuel level in the tank.

PROPELLANT MANAGEMENT

The propellant management systems provide a means of monitoring and controlling propellants during all phases of stage operation. Continuous capacitance probes and point level sensors in the LH$_2$ and lox tanks monitor propellant mass. During the propellant loading sequence, the point level sensors are used to indicate to GSE the level of propellants in the tanks. In flight, the level sensors provide signals to the LVDC in order to accomplish a smooth engine cutoff at propellant depletion. The capacitance probes provide outputs which are telemetered to ground stations so that propellant consumption can be monitored and recorded. Figure 5-14 and 5-15 illustrate the components of the systems.

PROPELLANT UTILIZATION SUBSYSTEM

The propellant utilization (PU) subsystem provides the capability to control the mixture ratio (MR) of lox to LH$_2$. The PU subsystem consists of a rotary valve which controls the amount of lox flowing to the engine, and electrical controls for the valve. At the time of J-2 engine start signal the PU valve is in a neutral position which provides a MR of

approximately 5.0:1. Approximately 5.5 seconds after J-2 engine start the LVDC in the IU commands the PU valve to a position which will provide a MR of approximately 5.5:1. The LVDC in the IU will again command the PU valve control circuitry when the LV has attained a preprogrammed velocity increase. This will occur approximately 5 minutes 25 seconds after J-2 engine start and will provide a MR of approximately 4.3:1 for the remainder of the burn. The excursion effect caused by varying the MR is illustrated in figure 5-8. Figure 2-26 illustrates the times of MR shifts on a thrust versus time curve. For a representative block diagram of the PU subsystem see figure 6-21.

Propellant Depletion

Five discrete liquid level sensors in each propellant tank provide initiation of engine cutoff upon detection of propellant depletion. The LH$_2$ tank sensors are located above each feedline outlet while the lox tank sensors are located directly above the sump. The cutoff sensors will initiate a signal to shutdown the engines when two out of five engine cutoff signals from the same tank are received.

ELECTRICAL

The electrical system is comprised of the electrical power and electrical control subsystems. The electrical power system provides the S-II stage with the electrical power source and distribution. The electrical control system interfaces with the IU to accomplish the mission requirements of the stage. The LVDC in the IU controls inflight sequencing of stage functions through the stage switch selector. The stage switch selector can provide up to 112 individual outputs in response to the appropriate commands. These outputs are routed through the stage electrical sequence controller or the separation controller to accomplish the directed operation. These units are basically a network of low power transistorized switches that can be controlled individually and, upon command from the switch selector, provide properly sequenced electrical signals to control the stage functions.

ELECTRICAL POWER SYSTEM

The electrical power system consists of six dc bus systems and a ground supplied ac bus system. In flight the electrical power system busses are energized by four zinc-silver oxide batteries. See figure 5-16 for battery characteristics. An integral heater and temperature probe are included in each battery. Power for battery heaters and for auxiliary hydraulic pump motors is supplied by GSE and is available only during prelaunch operations. Stage-mounted motor driven power transfer switches are employed to remotely disconnect all batteries from busses until just before launch. Approximately 50 seconds prior to liftoff, a power transfer sequence is initiated which changes the source power over to the stage mounted batteries. During the prelaunch checkout period all electrical power is supplied from GSE.

The motorized power transfer switches have a make-before-break (MBB) action to prevent power interruption during transfer from ground power to onboard battery power.

Each power source has an independant distribution system. There are no provisions for switching between the primary power sources or their associated distribution systems. No electrical failure of any type in one system can cause a failure in the other systems.

S-II BATTERY CHARACTERISTICS	
Type	Dry charge
Material	Alkaline silver-zinc
Electrolyte	Potassium hydroxide (KOH) in demineralized water
Cells	20 with taps to reduce voltage as required
Nominal voltage	28 ± 2 vdc
Current rating	35 amp hours
Gross weight	165 pounds

Figure 5-16

Distribution

Figure 5-17 illustrates the electrical system distribution. The loads have been distributed between the various busses in accordance with the following criteria:

1. Inflight loads, critical to mission continuance without performance degradation, are supplied by the main dc bus.

2. All instrumentation loads are supplied by the instrumentation dc bus.

3. All loads operational only on the ground are isolated from flight loads and supplied from ground power.

4. Two independent power sources supply the propellant dispersion, emergency detection systems and separation systems.

5. The recirculation pump motor system is supplied from a 56 volt dc system (two 28 volt batteries in series).

The division of loads between the main dc bus and the instrumentation dc bus leads to several advantages: closer voltage regulation and freedom from voltage variations is obtained; the number of loads on the main dc bus may be minimized and thus potential failure modes for the bus system minimized; instrumentation of most stage systems is

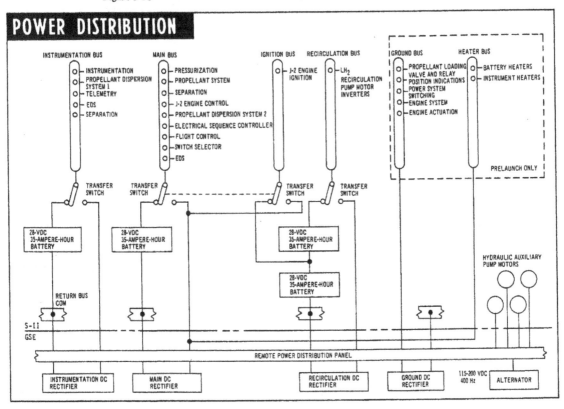

Figure 5-17

still maintained after partial or total failure of the main dc bus system so that failure analysis capability is maintained.

Primary power is fed to high current capacity busses in the power distributor. Power is then routed to auxiliary and control distributors, or to measuring distributors for instrumentation power. Components which require high current levels are supplied directly from the main power distributor busses.

ELECTRICAL CONTROL SYSTEM

The electrical control system provides electrical control for the various stage-mounted systems to implement normal flight operations, or prelaunch operations and checkout functions. The electrical control system contains most of the electrical and electronic components that are installed on the stage and required by the various mechanical systems for normal stage operation.

The primary stimuli for the electrical control system are provided by the IU and/or the GSE. Through the switch selector and sequence controller, various subsystems and functions are controlled. They include:

1. The propellant feed system which controls the fill and drain valves and the recirculation and conditioning valves.

2. The propellant management system which helps control the fill operation and the propellant utilization system during flight.

3. The pressurization system which controls the tank prepressurization valves, inflight pressurization valves, and the vent/relief valves.

4. The separation system which provides for S-IC/S-II separation, S-II ullage rocket ignition, S-II skirt separation, S-II/S-IVB separation, and S-II retrorocket ignition.

5. The propellant dispersion system which provides for thrust termination through engine cutoff and for explosively rupturing the propellant tanks.

6. The Emergency Detection System.

7. The starting and cutoff of the J-2 engine and the monitoring of certain engine system conditions.

8. The operation of the propellant prevalves.

INSTRUMENTATION

The S-II instrumentation system consists of an operational measurements and telemetry system. The measurement system monitors and measures conditions on the S-II stage while the telemetry system transmits this information to ground stations.

MEASUREMENT SYSTEM

The measurement system consists of transducers, signal conditioners, and distribution equipment necessary to provide the required measurement ranges and to present suitably scaled signals to the telemetry system.

The measurement system monitors numerous stage conditions and characteristics. This data is processed and conditioned into a form acceptable to the telemetry systems.

Measurements fall into a number of basic categories depending upon the type of measured variable, the variable rate of change with time, and other considerations. Because the stage engines are ignited in flight, a large number of engine and environmental control measurements are required. Measurement parameters monitored include acceleration, acoustics, discretes, flowrates, liquid levels, positions, pressures, RPM, strain, temperature, vibration, voltage, current, frequency, etc.

TELEMETRY SYSTEM

The telemetry system accepts the signals produced by the measuring portion of the instrumentation system and transmits them to the ground stations. Telemetry equipment includes signal multiplexers, subcarrier oscillators, amplifiers, modulators, transmitters, RF power amplifiers, RF multiplexers and an omnidirectional system of four antennae. The telemetry subsystem uses multiplex techniques (signal mixing and time sharing) to transmit large quantities of measurement data over a relatively small number of basic RF links (figure 5-18).

Inflight data is transmitted in the form of frequency-modulated RF carriers in the 230 to 250-MHz band, through the common omnidirectional antenna system.

ANTENNAE

Four antennae, installed at 90 degree intervals (see figure 5-2), are employed to provide omnidirectional coverage. The antennae are linear, cavity-backed, slot antennae which are fed from a hybrid junction ring and power dividers.

ORDNANCE

The S-II ordnance systems include the separation, ullage rocket, retrorocket, and propellant dispersion (flight termination) systems.

SEPARATION SYSTEM

The Saturn V launch vehicle system provides for separation of an expended stage from the remainder of the vehicle. For S-IC/S-II separation, a dual plane separation technique is used wherein the structure between the two stages is severed at two different planes (figure 5-19). The S-II/S-IVB separation occurs at a single plane (figure 5-19). . All separations are controlled by the launch vehicle digital computer (LVDC) located in the IU.

A sequence of events for S-IC/S-II/S-IVB separations and a block diagram of the separation systems is shown in figure 5-20.

Ordnance for first plane separation consists of two exploding bridgewire (EBW) firing units, two EBW detonators, and one linear shaped charge (LSC) assembly, which includes the LSC (containing 25 grains per foot of RDX) with a detonator block on each end (figures 5-19 and 5-20). The EBW firing units are installed on the S-IC/S-II interstage slightly below the S-II first separation plane. The leads of the EBW firing units are attached to the EBW detonators which are installed in the detonator blocks of the LSC assembly. The LSC detonator blocks are installed on adjustable mounts to provide for length variations of the LSC assembly and the circumference tolerances of the interstage. The LSC is routed from the detonator blocks around the periphery of the interstage.

The LSC is held in place by retaining clips and encased by covers which are secured by clips and sealed to environmentally protect the LSC. The two EBW firing units provide redundant signal paths for initiation of the LSC assembly. The storage capacitor in each of the EBW firing units is charged by 28 vdc power during the latter part of S-IC boost. The trigger signal causes the storage capacitor to discharge into an EBW detonator which explodes the bridgewire to release energy to detonate the explosive charge in the detonator. The output of the detonators initiates each

end of the LSC assembly.

Detonation of the LSC assembly severs the tension members attaching the S-IC/S-II interstage at station 1564.

The second plane separation ordnance is similar in composition and function to that of the first plane separation. The EBW firing units are installed on the S-IC/S-II interstage slightly below the separation plane. Detonation of the LSC assembly severs the tension members attaching the S-IC/S-II interstage at station 1760.

No heat-sensitive primary explosives are used and the detonators are not sensitive to accidental application of vehicle or gound power, static discharge, or RF energy. A spark gap in one pin of the firing circuitry prevents burnout of the bridgewire if power is accidentally applied.

S-II/S-IVB third plane separation is discussed in Section VI.

Figure 5-18

STAGE SEPARATION SYSTEMS

Figure 5-19

S-24

ULLAGE ROCKET SYSTEM

To ensure stable flow of propellants into the J-2 engines, a small forward acceleration is required to settle the propellants in their tanks. This acceleration is provided by ullage rockets (figures 5-19 and 5-21).

The S-II ullage rocket system consists of two EBW firing units, two EBW detonators, two CDF manifolds, nine CDF assemblies, eight CDF initiators and four ullage rockets. CDF assemblies connect the two CDF manifolds together and both manifolds to each of the four ullage rockets (see block diagram on figure 5-20). The ullage rockets are mounted parallel to vehicle centerline, 90 degrees apart on the periphery of the S-IC/S-II interstage at its aft end (figure 5-19). The rocket nozzles are just above the first separation plane and are canted outward 10 degrees to reduce the moment that would result from one or more rockets malfunctioning and to reduce exhaust plume impingement. With any one ullage rocket inoperative, the remaining rockets are capable of maintaining a minimum vehicle acceleration necessary for proper S-II engine ignition.

Each ullage rocket contains approximately 336 pounds of solid propellant, cast-in-place, in a four point star

configuration. Ammonium perchlorate composes 82 percent of the propellant weight. The case is 4130 steel. The rocket is approximately 89 inches long by 12-1/2 inches in diameter and develops 23,000 pounds of thrust in a burn time of 3.75 seconds.

RETROROCKET SYSTEM

To separate and retard the S-II stage, a deceleration is provided by the retrorocket system.

The system consists of two EBW firing units, two EBW detonators, two CDF manifolds, nine CDF assemblies, eight pyrogen initiators, and four retrorockets (figure 5-21). The components are connected to each other in a manner similar to that of the ullage rocket system (see block diagram on figure 5-20). The retrorockets are mounted 90 degrees apart in the aft end of S-II/S-IVB interstage between stations 2519 and 2633 (figure 5-21). The retrorockets are canted out from the vehicle centerline approximately three degrees with the nozzles canted out nine and one-half degrees from the centerline.

Each retrorocket contains approximately 268.2 pounds of case-bonded, single-grain, solid propellant with a tapered,

S-IC/S-II AND S-II/S-IVB SEPARATION

S-IC/S-II separation

 1. EBW firing units enabled

A ground-latched interlock renders all the EBW firing units on the Saturn V inoperative while the vehicle is on the launch pad. The interlock is released with umbilical disconnect during liftoff, and the subsystem is reset to flight conditions.

2. S-IC/S-II separation ordnance arm

The ordnance-arm command is routed through the S-II switch selector to both the S-IC stage electrical circuitry to supply 28 vdc to the EBW units for first-plane separation and retrorocket ignition, and to the S-II stage electrical circuitry to supply 28 vdc to the EBW units for ullage rocket ignition and second-plane separation.

 3. S-IC outboard engine cutoff followed by S-II ullage rocket ignition

 4. First-plane separation

Second-plane separation is enabled by the removal of an electrical interlock during first-plane separation.

 5. Second-plane separation

The second-plane separation command is generated by the IU approximately thirty seconds after first-plane separation.

This delay permits the transient vehicle motion, associated with first-plane separation, to dampen out.

The separation command is routed to the S-II switch selector to trigger the ordnance train and ignite the LSC for second-plane separation. The LSC detonates, severing the S-II interstage from the S-II stage. The combined effect of vehicle acceleration and the reaction caused by the J-2 engine exhaust plume impingement retards the interstage .

S-II/S-IVB separation

Physical separation is initiated by the IU at the end of the S-II boost phase following shutdown of the five J-2 engines. Separation requires the performance of the following major functions in the sequence described:

 6. S-II/S-IVB separation ordnance arm

The ordnance-arm command is routed through the S-II switch selector to both the S-II and S-IVB stage electrical circuitry and carries 28 vdc to the EBW firing units for S-II/S-IVB separation and retrorocket ignition.

 7. S-II/S-IVB separation

Four solid propellant S-II retrockets (figure 5-21) are mounted at equal intervals on the periphery of the S-II/S-IVB interstage structure and are used to retard the S-II stage after separation.

Figure 5-20 (Sheet 1 of 2)

five-point star configuration. The 4130 steel case is 9 inches in diameter and 90.68 inches long. The approximate length and weight of the rocket are 104.68 inches and 377.5 pounds, respectively. Each produces a thrust of 34,810 pounds in 1.52 seconds of burning time.

PROPELLANT DISPERSION SYSTEM

The S-II propellant dispersion system (PDS) provides for termination of vehicle flight during the S-II boost phase if the vehicle flight path varies beyond its prescribed limits or if continuation of vehicle flight creates a safety hazard. The S-II PDS may be safed after the launch escape tower is jettisoned.

The system is installed in compliance with Air Force Eastern Test Range (AFETR) Regulation 127-9 and AFETR Safety Manual 127-1.

The S-II PDS is a dual channel, redundant system composed of two segments (figure 5-22). The radio frequency segment receives, decodes, and controls the propellant dispersion commands. The ordnance train segment consists of two EBW firing units, two EBW detonators, one safety and arming (S&A) device (shared by both channels), six CDF assemblies, two CDF tees, one LH$_2$ tank LSC assembly, two lox tank destruct charge adapters and one lox tank destruct charge assembly.

Figure 5-20 (Sheet 2 of 2)

S-II ULLAGE AND RETROROCKETS

CDF ASSEMBLY

FOR PRESSURE MEASUREMENT

CDF ASSEMBLY

CDF INITIATOR

CDF INITIATOR

IGNITER ADAPTER

PLUG

JETTISONABLE FAIRING ASSEMBLY

FIXED FAIRING

S-IVB INTERSTAGE ASSEMBLY

S-II/S-IVB FIELD SPLICE

ADJUSTABLE ROD ASSEMBLIES

CDF INITIATORS

CDF ASSEMBLIES

CAP

S-II INTERSTAGE ASSEMBLY

FIRST SEPARATION PLANE

FAIRING

TYPICAL S-II ULLAGE ROCKET (4 REQUIRED)

TYPICAL S-II RETROROCKET (4 REQUIRED)

Figure 5-24

PROPELLANT DISPERSION

Figure 5-22

Should emergency flight termination become necessary, two coded radio frequency commands are transmitted to the launch vehicle by the range safety officer. The first command arms the EBW firing units (figure 5-22) and initiates S-II stage engine cutoff. The second command, which is delayed to permit charging of the EBW firing units, discharges the storage capacitors in the EBW firing units across the exploding bridgewire in the EBW detonators mounted on the S&A device. The resulting explosive wave propagates through the S&A device inserts to the CDF assemblies and to the CDF tees. The CDF tees, installed on the S-II forward skirt, propagate the wave to two CDF assemblies which detonate to their respective destruct assemblies. The destruct assemblies are connected by a CDF assembly to provide redundancy to the system.

A description of the S&A device is included in the PDS discussion in Section IV.

The LH_2 tank linear shaped charge, when detonated, cuts a 30-foot vertical opening in the tank. The LSC assembly consists of two 15-foot sections of RDX loaded at 600 grains per foot.

The lox tank destruct charges cut 13-foot lateral openings in the lox tank and the S-II aft skirt simultaneously. The destruct assembly consists of two linear explosive charges of RDX loaded at 800 grains per foot. The destruct charges are installed in a figure-eight tube mounted on the inside of the aft skirt structure near station number 183.0.

MAJOR DIFFERENCES

There are no major differences between the S-II stages on vehicles SA-506, SA-507, SA-508 and SA-509.

This Space Intentionally Left Blank

TABLE OF CONTENTS

INTRODUCTION

The Saturn S-IVB (figure 6-1) is the third booster stage. Its single J-2 engine is designed to boost the payload into a circular orbit on the first burn, then boost the payload to a proper position and velocity for lunar intercept with a second burn. The stage and engine characteristics are listed in figure 1-3. The major systems of the stage are: structures, environmental control, propulsion, propellants, pneumatic control, flight control, auxiliary propulsion system, electrical, instrumentation and ordnance.

STRUCTURE

The basic S-IVB stage airframe, illustrated in figure 6-1, consists of the following structural assemblies: the forward skirt, propellant tanks, aft skirt, thrust structure, and aft interstage. These assemblies, with the exception of the propellant tanks, are all of a skin/stringer type aluminum alloy airframe construction. In addition, there are two longitudinal tunnels which house wiring, pressurization lines, and propellant dispersion systems. The tunnel covers are made of aluminum stiffened by internal ribs. These structures do not transmit primary shell loads but act only as fairings.

FORWARD SKIRT ASSEMBLY

Cylindrical in shape, the forward skirt (figure 6-1) extends forward from the intersection of the liquid hydrogen (LH_2) tank sidewall and the forward dome, providing a hard attach point for the Instrument Unit (IU). It is the load supporting member between the LH_2 tank and the IU. An access door in the IU allows servicing of the equipment in the forward skirt. The five environmental plates which support and thermally condition various electronic components, such as the transmitters and signal conditioning modules, are attached to the inside of this skirt. The forward umbilical plate, antennae, LH_2 tank flight vents and the tunnel fairings are attached externally to this skirt.

PROPELLANT TANK ASSEMBLY

The propellant tank assembly (figure 6-1) consists of a cylindrical tank with a hemispherical shaped dome at each end, and a common bulkhead to separate the lox from the LH_2. This bulkhead is of sandwich type construction consisting of two parallel, hemispherical shaped, aluminum

alloy (2014-T6) domes bonded to a fiberglass-phenolic honeycomb core. The internal surface of the LH_2 tank is machine milled in a waffle pattern to obtain required tank stiffness with minimum structural weight. To minimize LH_2 boil off, polyurethane insulation blocks, covered with a fiberglass sheet and coated with a sealant, are bonded into the milled areas of the waffle patterns.

The walls of the tank support all loads forward of the forward skirt attach point and transmit the thrust to the payload. Attached to the inside of the LH_2 tank are; a 34 foot continuous capacitance probe, nine cold helium spheres, brackets with temperature and level sensors, a chilldown pump, a slosh baffle, a slosh deflector, and fill, pressurization and vent pipes. Attached to the inside of the lox tank are; slosh baffles, a chilldown pump, a 13.5 foot continuous capacitance probe, temperature and level sensors, and fill, pressurization and vent pipes. Attached externally to the propellant tank are helium pipes, propellant dispersion components, and wiring which passes through two tunnel fairings. The forward edge of the thrust structure is attached to the lox tank portion of the propellant tank.

THRUST STRUCTURE

The thrust structure assembly (figure 6-1) is an inverted, truncated cone attached at its large end to the aft dome of the lox tank and attached at its small end to the engine mount. It provides the attach point for the J-2 engine and distributes the engine thrust over the entire tank circumference. Attached external to the thrust structure are the engine piping, wiring and interface panels, eight ambient helium spheres, hydraulic system, oxygen/hydrogen burner, and some of the engine and lox tank instrumentation.

AFT SKIRT ASSEMBLY

The cylindrical shaped aft skirt assembly is the load bearing structure between the LH_2 tank and aft interstage. The aft skirt assembly is bolted to the tank assembly at its forward edge and connected to the aft interstage. A frangible tension tie separates it from the aft interstage at S-II separation.

AFT INTERSTAGE ASSEMBLY

The aft interstage is a truncated cone that provides the load supporting structure between the S-IVB stage and the S-II stage (figure 6-1). The interstage also provides the focal point for the required electrical and mechanical interface between the S-II and S-IVB stages. The S-II retrorocket motors are attached to this interstage and at separation the interstage remains attached to the S-II stage.

ENVIRONMENTAL CONTROL

There are three general requirements for environmental control during checkout and flight operations of the S-IVB stage. The first is associated with ground checkout and prelaunch operations and involves thermal conditioning of the environment around the electrical equipment, auxiliary propulsion system (APS), and hydraulic accumulator

S-IVB STAGE STRUCTURE

FORWARD SKIRT

10.2 FEET

21.6 FEET

PROPELLANT TANK

LH$_2$ TANK

10,418 CU FT

44.0 FEET

LOX TANK

2,830 CU FT

59.0 FEET

7.0 FEET

AFT SKIRT

THRUST STRUCTURE (WITH ENGINE ATTACHED)

5.2 FEET

33.0 FEET

19 FEET

AFT INTERSTAGE

Figure 6-1

reservoir. In addition, there is a requirement for aft skirt and interstage purging. The second involves forward skirt area purging, while the third concerns inflight heat dissipation for the electrical/electronic equipment.

AFT SKIRT AND INTERSTAGE THERMOCONDITIONING

During countdown, air/GN_2 is supplied by the environmental control system, which is capable of switching from air to GN_2 purge. Air or GN_2 is supplied at the rate of approximately 3600 scfm. The air purge is initiated when electrical power is applied to the vehicle. GN_2 flow is initiated approximately two hours prior to cryogenic loading and continued until liftoff. During periods of hold, GN_2 purge is continued. The aft skirt and interstage thermoconditioning and purge subsystem provides the following:

1. Thermal conditioning of the atmosphere around electrical equipment in the aft skirt during ground operations.

2. Thermal conditioning of the APS, hydraulic accumulator reservoir, and control system helium sphere.

3. Purging of the aft skirt, aft interstage and thrust structure, and the forward skirt of the S-II stage of oxygen, moisture and combustible gases.

The subsystem consists of a temperature-controlled air or GN_2 distribution system (figure 6-2). The purging gas passes over electrical equipment below the ring frame and flows into the interstage. A duct from the skirt manifold directs air or GN_2 to a thrust structure manifold. Another duct directs the gas to a shroud covering the control system helium sphere which is used to purge the lox and LH_2 pump shaft seal cavities. From the thrust structure manifold supply duct, a portion of air or GN_2 is directed to a shroud covering the hydraulic accumulator reservoir.

Temperature control is accomplished by two dual element thermistor assemblies located in the gaseous exhaust stream of each of the APS modules. The elements are wired such that each single element in one thermistor assembly is wired in series with a single element in the other thermistor assembly in order to provide two sets of average temperature indications. One set is used for temperature control, the other for temperature recording.

FORWARD SKIRT THERMOCONDITIONING

The electrical/electronic equipment in the S-IVB forward skirt area is thermally conditioned by a heat transfer subsystem using a circulating coolant for the medium. Principal components of the system, located in the S-IVB stage forward skirt area, are a fluid distribution subsystem and cold plates. The coolant is supplied to the S-IVB by the IU thermoconditioning system starting when electrical power

Figure 6-2

is applied to the vehicle and continuing throughout the mission. For a description of this system refer to Section VII.

FORWARD SKIRT AREA PURGE

The forward skirt area is purged with GN_2 to minimize the danger of fire and explosion while propellants are being loaded or stored in the stage, or during other hazardous conditions. The purge is supplied by the IU purge system which purges the entire forward skirt/IU/adapter area. The total flow rate into this area is approximately 3500 scfm.

PROPULSION

This stage provides vehicle propulsion twice during the mission. The first burn occurs immediately after S-II/S-IVB separation and lasts long enough to insert the vehicle into earth orbit. The second burn injects the S-IVB/IU/spacecraft into a translunar trajectory.

At J-2 engine first burn cutoff the auxiliary propulsion system (APS) ullage engines are ignited and burn about one and one half minutes providing stabilization and settling of the propellants. At APS engine ignition, the APS yaw and pitch control modes are enabled (roll already active) for the required attitude control of the stage during coast. LH_2 continuous venting is activated about one minute after first burn cutoff and continues until oxygen/hydrogen (O_2/H_2) burner start for second burn.

Prior to second burn, the systems are again readied for an engine start. Approximately five minutes before restart, the chilldown systems are reactivated to condition the lines by removing gases collected in the propellant supply system. The O_2/H_2 burner is started approximately nine minutes prior to second burn and operates to settle the propellants and pressurize the propellant tanks until approximately one minute before engine start, when the APS ullage engines are fired, the O_2/H_2 burner is shut down, and ambient repressurization is started. Then the chilldown systems are deactivated, and engine restart is initiated. The APS ullage engines are shut off and ambient repressurization is stopped as the engine reaches full thrust. The APS yaw and pitch control modes are deenergized (roll control mode remains active) during the second burn. Following the second burn, the transportation and docking maneuver and final separation of the spacecraft from the launch vehicle are accomplished. The S-IVB stage is then placed into solar orbit by dumping residual propellants through the engine.

J-2 ROCKET ENGINE

The J-2 rocket engine (figure 5-6) is a high performance, multiple restart engine utilizing liquid oxygen and liquid hydrogen as propellants. The engine attains a thrust of 204,000 pounds during both burns. The only substances used in the engine are the propellants and helium gas. The extremely low operating temperature of the engine prohibits the use of lubricants or other fluids. The engine features a single, tubular-walled, bell-shaped thrust chamber and two independently driven, direct drive, turbopumps for liquid oxygen and liquid hydrogen. Both turbopumps are powered in series by a single gas generator, which utilizes the same propellants as the thrust chamber. The main hydraulic pump is driven by the oxidizer turbopump turbine. The ratio of fuel to oxidizer is controlled by bypassing liquid oxygen

from the discharge side of the oxidizer turbopump to the inlet side through a servovalve.

The engine valves are controlled by a pneumatic system powered by gaseous helium which is stored in a sphere inside the start bottle. An electrical control system, which uses solid state logic elements, is used to sequence the start and shutdown operations of the engine. Electrical power is supplied from aft battery No. 1.

During the burn periods, the lox tank is pressurized by flowing cold helium through the heat exchanger in the oxidizer turbine exhaust duct. The heat exchanger heats the cold helium, causing it to expand. The LH_2 tank is pressurized during burn periods by GH_2 from the thrust chamber fuel manifold.

During burn periods, control in the pitch and yaw planes, is achieved by gimbaling the main engine. Hydraulic pressure for gimbal actuation is provided by the main hydraulic pump. During coast mode the APS engines give the pitch and yaw thrust vector control. Roll control during both the burn periods and the coast modes is achieved by firing the APS engines.

ENGINE OPERATION

Engine operation includes starting, mainstage and cutoff. The starting and cutoff phases are periods of transition in which a sequence of activities occur. These phases are developed in detail in the following paragraphs.

Start Preparations

Preparations for an engine start include ascertaining the positions and status of various engine and stage systems and components. The J-2 engine electrical control system controls engine operation by means of electrical signals. The heart of the engine electrical control system is the electrical control package (17, figure 5-6). It sequences and times the functions required during engine start or cutoff.

Each cutoff automatically causes the electrical control package circuitry to reset itself, ready for restart, providing that all reset conditions are met. The LVDC issues an engine ready bypass signal just prior to each engine start attempt. This bypass signal acts in the same manner as a cutoff would act. The reset signals engine ready and this allows the LVDC to send its start command. Receipt of the start command initiates the engine start sequence.

Engine Start Sequence

When engine start is initiated (3, figure 6-3), the spark exciters in the electrical control package provide energy for the gas generator (GG) and augmented spark igniter (ASI) spark plugs (4). The helium control and ignition phase control valves, in the pneumatic control package (1), are simultaneously energized, allowing helium from the helium tank (2) to flow through the pneumatic regulator to the pneumatic control system. The helium is routed through the internal check valve in the pneumatic control package (1) to ensure continued pressure to the engine valves in the event of helium supply failure. The regulated helium fills a pneumatic accumulator, closes the propellant bleed valves (5), and purges (6) the oxidizer dome and gas generator oxidizer injector manifold. The oxidizer turbopump (12) intermediate reservoir. In addition, there is a requirement for aft skirt and interstage purging. The second involves forward skirt area purging, while the third concerns inflight heat dissipation for the electrical/electronic equipment.

AFT SKIRT AND INTERSTAGE THERMOCONDITIONING

During countdown, air/GN_2 is supplied by the environmental control system, which is capable of switching from air to GN_2 purge. Air or GN_2 is supplied at the rate of approximately 3600 scfm. The air purge is initiated when electrical power is applied to the vehicle. GN_2 flow is initiated approximately two hours prior to cryogenic loading and continued until liftoff. During periods of hold, GN_2 purge is continued. The aft skirt and interstage thermoconditioning and purge subsystem provides the following:

1. Thermal conditioning of the atmosphere around electrical equipment in the aft skirt during ground operations.

2. Thermal conditioning of the APS, hydraulic accumulator reservoir, and control system helium sphere.

3. Purging of the aft skirt, aft interstage and thrust structure, and the forward skirt of the S-II stage of oxygen, moisture and combustible gases.

The subsystem consists of a temperature-controlled air or GN_2 distribution system (figure 6-2). The purging gas passes over electrical equipment below the ring frame and flows into the interstage. A duct from the skirt manifold directs air or GN_2 to a thrust structure manifold. Another duct directs the gas to a shroud covering the control system helium sphere which is used to purge the lox and LH_2 pump shaft seal cavities. From the thrust structure manifold supply duct, a portion of air or GN_2 is directed to a shroud covering the hydraulic accumulator reservoir.

Temperature control is accomplished by two dual element thermistor assemblies located in the gaseous exhaust stream of each of the APS modules. The elements are wired such that each single element in one thermistor assembly is wired in series with a single element in the other thermistor assembly in order to provide two sets of average temperature indications. One set is used for temperature control, the other for temperature recording.

FORWARD SKIRT THERMOCONDITIONING

The electrical/electronic equipment in the S-IVB forward skirt area is thermally conditioned by a heat transfer subsystem using a circulating coolant for the medium. Principal components of the system, located in the S-IVB stage forward skirt area, are a fluid distribution subsystem and cold plates. The coolant is supplied to the S-IVB by the IU thermoconditioning system starting when electrical power

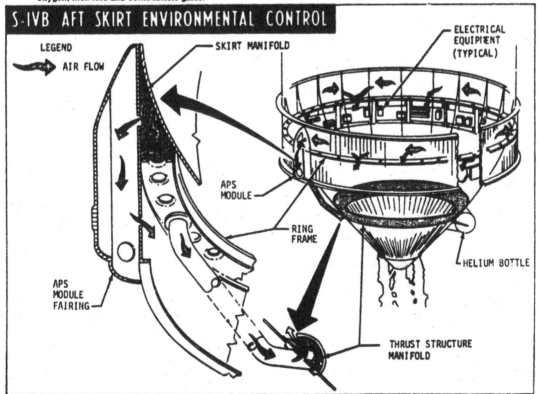

S-IVB AFT SKIRT ENVIRONMENTAL CONTROL

LEGEND

AIR FLOW

SKIRT MANIFOLD

ELECTRICAL EQUIPMENT (TYPICAL)

APS MODULE

RING FRAME

HELIUM BOTTLE

APS MODULE FAIRING

THRUST STRUCTURE MANIFOLD

Figure 6-2

is applied to the vehicle and continuing throughout the mission. For a description of this system refer to Section VII.

FORWARD SKIRT AREA PURGE

The forward skirt area is purged with GN_2 to minimize the danger of fire and explosion while propellants are being loaded or stored in the stage, or during other hazardous conditions. The purge is supplied by the IU purge system which purges the entire forward skirt/IU/adapter area. The total flow rate into this area is approximately 3500 scfm.

PROPULSION

This stage provides vehicle propulsion twice during the mission. The first burn occurs immediately after S-II/S-IVB separation and lasts long enough to insert the vehicle into earth orbit. The second burn injects the S-IVB/IU/spacecraft into a translunar trajectory.

At J-2 engine first burn cutoff the auxiliary propulsion system (APS) ullage engines are ignited and burn about one and one half minutes providing stabilization and settling of the propellants. At APS engine ignition, the APS yaw and pitch control modes are enabled (roll already active) for the required attitude control of the stage during coast. LH_2 continuous venting is activated about one minute after first burn cutoff and continues until oxygen/hydrogen (O_2/H_2) burner start for second burn.

Prior to second burn, the systems are again readied for an engine start. Approximately five minutes before restart, the chilldown systems are reactivated to condition the lines by removing gases collected in the propellant supply system. The O_2/H_2 burner is started approximately nine minutes prior to second burn and operates to settle the propellants and pressurize the propellant tanks until approximately one minute before engine start, when the APS ullage engines are fired, the O_2/H_2 burner is shut down, and ambient repressurization is started. Then the chilldown systems are deactivated, and engine restart is initiated. The APS ullage engines are shut off and ambient repressurization is stopped as the engine reaches full thrust. The APS yaw and pitch control modes are deenergized (roll control mode remains active) during the second burn. Following the second burn, the transportation and docking maneuver and final separation of the spacecraft from the launch vehicle are accomplished. The S-IVB stage is then placed into solar orbit by dumping residual propellants through the engine.

J-2 ROCKET ENGINE

The J-2 rocket engine (figure 5-6) is a high performance, multiple restart engine utilizing liquid oxygen and liquid hydrogen as propellants. The engine attains a thrust of 204,000 pounds during both burns. The only substances used in the engine are the propellants and helium gas. The extremely low operating temperature of the engine prohibits the use of lubricants or other fluids. The engine features a single, tubular-walled, bell-shaped thrust chamber and two independently driven, direct drive, turbopumps for liquid oxygen and liquid hydrogen. Both turbopumps are powered in series by a single gas generator, which utilizes the same propellants as the thrust chamber. The main hydraulic pump is driven by the oxidizer turbopump turbine. The ratio of fuel to oxidizer is controlled by bypassing liquid oxygen

from the discharge side of the oxidizer turbopump to the inlet side through a servovalve.

The engine valves are controlled by a pneumatic system powered by gaseous helium which is stored in a sphere inside the start bottle. An electrical control system, which uses solid state logic elements, is used to sequence the start and shutdown operations of the engine. Electrical power is supplied from aft battery No. 1.

During the burn periods, the lox tank is pressurized by flowing cold helium through the heat exchanger in the oxidizer turbine exhaust duct. The heat exchanger heats the cold helium, causing it to expand. The LH_2 tank is pressurized during burn periods by GH_2 from the thrust chamber fuel manifold.

During burn periods, control in the pitch and yaw planes, is achieved by gimbaling the main engine. Hydraulic pressure for gimbal actuation is provided by the main hydraulic pump. During coast mode the APS engines give the pitch and yaw thrust vector control. Roll control during both the burn periods and the coast modes is achieved by firing the APS engines.

ENGINE OPERATION

Engine operation includes starting, mainstage and cutoff. The starting and cutoff phases are periods of transition in which a sequence of activities occur. These phases are developed in detail in the following paragraphs.

Start Preparations

Preparations for an engine start include ascertaining the positions and status of various engine and stage systems and components. The J-2 engine electrical control system controls engine operation by means of electrical signals. The heart of the engine electrical control system is the electrical control package (17, figure 5-6). It sequences and times the functions required during engine start or cutoff.

Each cutoff automatically causes the electrical control package circuitry to reset itself, ready for restart, providing that all reset conditions are met. The LVDC issues an engine ready bypass signal just prior to each engine start attempt. This bypass signal acts in the same manner as a cutoff would act. The reset signals engine ready and this allows the LVDC to send its start command. Receipt of the start command initiates the engine start sequence.

Engine Start Sequence

When engine start is initiated (3, figure 6-3), the spark exciters in the electrical control package provide energy for the gas generator (GG) and augmented spark igniter (ASI) spark plugs (4). The helium control and ignition phase control valves, in the pneumatic control package (1), are simultaneously energized, allowing helium from the helium tank (2) to flow through the pneumatic regulator to the pneumatic control system. The helium is routed through the internal check valve in the pneumatic control package (1) to ensure continued pressure to the engine valves in the event of helium supply failure. The regulated helium fills a pneumatic accumulator, close the propellant bleed valves (5) and purge (6) the oxidizer dome and gas generator oxidizer injector manifold. The oxidizer turbopump (7) not circulat-

6-4

Figure 6-3 (Sheet 1 of 3)

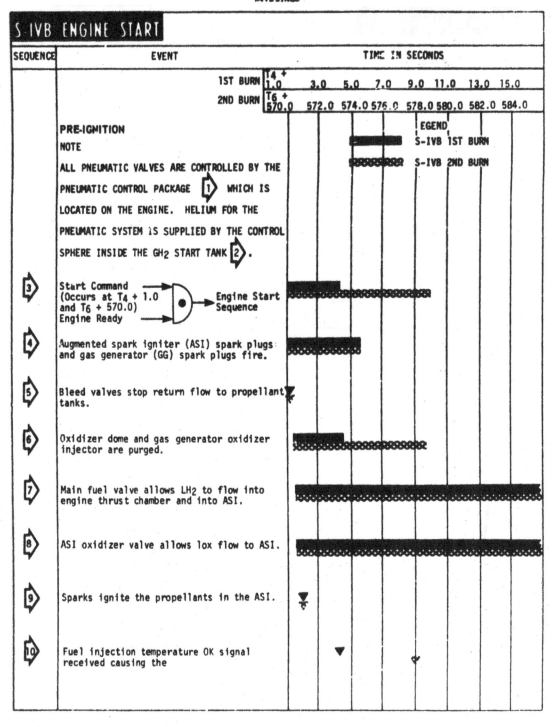

S-IVB ENGINE START

SEQUENCE	EVENT	TIME IN SECONDS

1ST BURN T4 + 1.0   3.0   5.0   7.0   9.0   11.0   13.0   15.0

2ND BURN T6 + 570.0   572.0   574.0   576.0   578.0   580.0   582.0   584.0

PRE-IGNITION

NOTE

ALL PNEUMATIC VALVES ARE CONTROLLED BY THE PNEUMATIC CONTROL PACKAGE ⟨1⟩ WHICH IS LOCATED ON THE ENGINE. HELIUM FOR THE PNEUMATIC SYSTEM IS SUPPLIED BY THE CONTROL SPHERE INSIDE THE GH_2 START TANK ⟨2⟩.

LEGEND
S-IVB 1ST BURN
S-IVB 2ND BURN

⟨3⟩ Start Command (Occurs at T4 + 1.0 and T6 + 570.0) Engine Ready → Engine Start Sequence

⟨4⟩ Augmented spark igniter (ASI) spark plugs and gas generator (GG) spark plugs fire.

⟨5⟩ Bleed valves stop return flow to propellant tanks.

⟨6⟩ Oxidizer dome and gas generator oxidizer injector are purged.

⟨7⟩ Main fuel valve allows LH_2 to flow into engine thrust chamber and into ASI.

⟨8⟩ ASI oxidizer valve allows lox flow to ASI.

⟨9⟩ Sparks ignite the propellants in the ASI.

⟨10⟩ Fuel injection temperature OK signal received causing the

Figure 6-3 (Sheet 2 of 3)

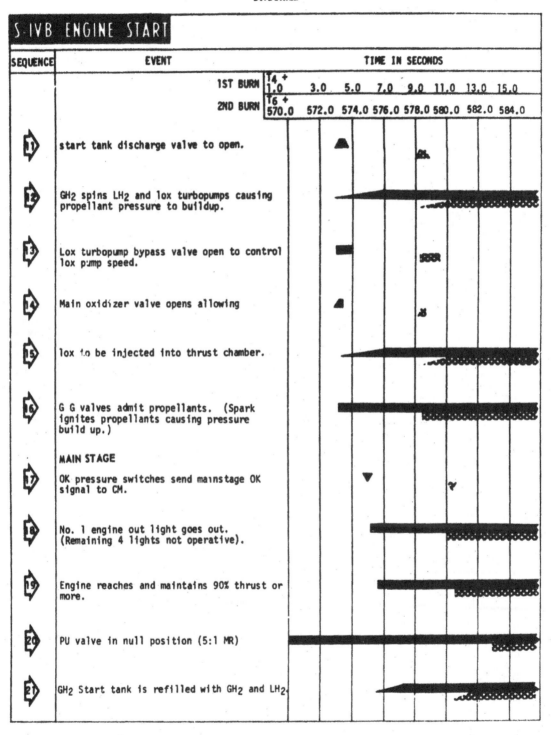

S-IVB ENGINE START

SEQUENCE	EVENT	TIME IN SECONDS							
	1ST BURN	T4 + 1.0	3.0	5.0	7.0	9.0	11.0	13.0	15.0
	2ND BURN	T6 + 570.0	572.0	574.0	576.0	578.0	580.0	582.0	584.0
11	start tank discharge valve to open.								
12	GH2 spins LH2 and lox turbopumps causing propellant pressure to buildup.								
13	Lox turbopump bypass valve open to control lox pump speed.								
14	Main oxidizer valve opens allowing								
15	lox to be injected into thrust chamber.								
16	G G valves admit propellants. (Spark ignites propellants causing pressure build up.)								
17	MAIN STAGE OK pressure switches send mainstage OK signal to CM.								
18	No. 1 engine out light goes out. (Remaining 4 lights not operative).								
19	Engine reaches and maintains 90% thrust or more.								
20	PU valve in null position (5:1 MR)								
21	GH2 Start tank is refilled with GH2 and LH2								

Figure 6-3 (Sheet 3 of 3)

seal cavity is continuously purged. The mainstage control valve holds the main oxidizer valve closed, and opens the purge control valve, which allows the oxidizer dome and gas generator oxidizer injector to be purged (6). The mainstage control valve also supplies opening control pressure to the oxidizer turbine bypass valve (13). The ignition phase control valve, when actuated, opens the main fuel valve (7) and the ASI oxidizer valve (8), and supplies pressure to the sequence valve located within the main oxidizer valve (14). Fuel is tapped from downstream of the main fuel valve for use in the ASI (4). Both propellants, under tank pressure, flow through the stationary turbopumps (12).

The sequence valve, in the main fuel valve (7), opens when the fuel valve reaches approximately 90% open, and routes helium to the start tank discharge valve (STDV) (11) control valve. Simultaneously with engine start, the STDV delay timer is energized. Upon expiration of the STDV timer and the receipt of a fuel injection temperature OK signal, the STDV control valve and ignition phase timer are energized. As the STDV control valve energizes, the discharge valve opens allowing gaseous hydrogen under pressure to flow through the series turbine drive system. This accelerates both turbopumps (12) to the proper operating levels to allow subsequent ignition and power buildup of the gas generator (16). The relationship of fuel-to-lox turbopump speed buildup is controlled by an orifice in the oxidizer turbine bypass valve (13). During the start sequence, the normally open oxidizer bypass valve (13) permits a percentage of the gas to bypass the oxidizer turbine.

During this period, ASI combustion occurs. Absence of a start tank depressurized signal will cause cutoff at the expiration of the ignition phase timer. Presence of the start tank depressurized signal allows the mainstage control valve to energize. Simultaneously, the sparks deenergize timer is energized and the STDV control valve is deenergized, causing the STDV to close. Helium pressure is vented from the main oxidizer valve (14) and from the purge control valve through the mainstage control valve. The purge control valve closes, terminating the oxidizer dome and gas generator oxidizer injector manifold purges (6). Pressure from the mainstage control valve is routed to open the main oxidizer valve (14). A sequence valve, operated by the main oxidizer valve (14), permits gaseous helium to open the gas generator control valve (4) and to close the oxidizer turbine bypass valve (13). Flow to close the oxidizer turbine bypass valve (13) is restricted as it passes through an orifice.

Propellants flowing into the gas generator (16) are ignited by the spark plugs (4). Combustion of the propellants causes the hot gases to drive the turbopumps (12). The turbopumps force propellant into the thrust chambers where it is ignited by the torch from the ASI.

Transition into mainstage occurs as the turbopumps (12) accelerate to steadystate speeds. As the oxidizer pressure increases, a thrust OK signal is generated by either of the two thrust OK pressure switches (17). (Cutoff occurs if no signal is received before expiration of the sparks deenergized timer.) The ASI and GG spark exciters are deenergized at expiration of the sparks deenergized timer. Cutoff occurs if both pressure switch actuated signals (thrust OK) are lost during mainstage operation.

Steadystate operation is maintained until a cutoff signal is initiated. During this period, gaseous hydrogen is tapped

from the fuel injection manifold to pressurize the LH$_2$ tank. The lox tank is pressurized by gaseous helium heated by the heat exchanger in the turbine exhaust duct. Gaseous hydrogen is bled from the thrust chamber fuel injection manifold, and liquid hydrogen is bled from the ASI fuel line to refill start tank for engine restart.

Engine Cutoff

The J-2 engine may receive cutoff signals from the following sources: EDS No.'s 1 and 2, range safety systems No.'s 1 and 2, thrust OK pressure switches, propellant depletion sensors, and an IU programmed command via the switch selector.

The switch selector, range safety system No. 2, EDS No. 2, and the propellant depletion sensors cutoff commands are tied together (but diode isolated) and sent to the electrical control package cutoff circuit. The dropout of the thrust OK pressure switches removes a cutoff inhibit function in the electrical control package cutoff circuit. EDS No. 1 and range safety system No. 1 cutoff commands will indirectly transfer the engine control power switch to the OFF position, causing the engine to shut down due to power loss.

Cutoff Sequence

The engine cutoff sequence is shown graphically in figure 6-4. When the electrical control package receives the cutoff signal (1), it deenergizes the mainstage and ignition phase control valves in the pneumatic control package (2), while energizing the helium control deenergize timer. The mainstage control valve closes the main oxidizer valve (3), and opens the purge control valve and the oxidizer turbine bypass valve (8). The purge control valve directs a helium purge (11) to the oxidizer dome and GG oxidizer injector. The ignition phase control valve closes the ASI oxidizer valve (4) and the main fuel valve (5), while opening the fast shutdown valve. The fast shutdown valve now rapidly vents the return flow from the GG control valve. All valves, except the ASI oxidizer valve (4) and oxidizer turbine bypass valve (8), are spring loaded closed. This causes the valves to start moving closed as soon as the pressure to open them is released. GG combustion pressure aids closing of the GG control valve.

Expiration of the helium control deenergize timer causes the helium control valve to close. When the helium control valve closes, it causes the oxidizer dome and GG oxidizer injector purges (11) to stop. An orifice in the locked up lines bleeds off pressure from the propellant bleed valves (13). This loss of pressure allows springs to open the valves. When open, the propellant bleed valves allow propellants to flow back to the propellant tanks.

Restart

The restart of the J-2 engine is identical to the initial start except for the fill procedure of the start tank. The start tank is filled with LH$_2$ and GH$_2$ during the previous burn period.

To ensure that sufficient energy will be available for spinning the LH$_2$ and lox pump turbines, a waiting period of approximately 80 minutes to 6 hours is permitted. The minimum time is required to build sufficient pressure by warming the start tank through natural means. The minimum wait is also needed to allow the hot gas turbine exhaust system to cool. Prolonged heating will cause a loss of energy in the start tank. This loss occurs when the LH$_2$ and GH$_2$

Figure 6-4 (Sheet 1 of 2)

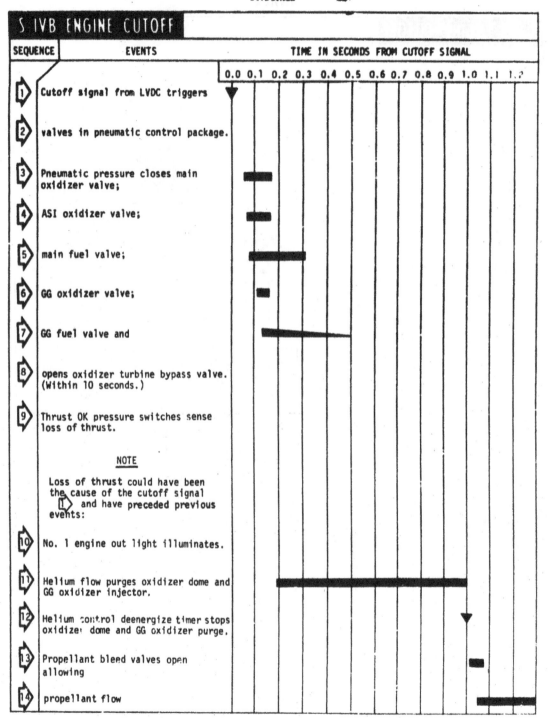

SEQUENCE	EVENTS	TIME IN SECONDS FROM CUTOFF SIGNAL

S IVB ENGINE CUTOFF

0.0 0.1 0.2 0.3 0.4 0.5 0.6 0.7 0.8 0.9 1.0 1.1 1.2

1 Cutoff signal from LVDC triggers

2 valves in pneumatic control package.

3 Pneumatic pressure closes main oxidizer valve;

4 ASI oxidizer valve;

5 main fuel valve;

6 GG oxidizer valve;

7 GG fuel valve and

8 opens oxidizer turbine bypass valve. (Within 10 seconds.)

9 Thrust OK pressure switches sense loss of thrust.

NOTE

Loss of thrust could have been the cause of the cutoff signal and have preceded previous events:

10 No. 1 engine out light illuminates.

11 Helium flow purges oxidizer dome and GG oxidizer injector.

12 Helium control deenergize timer stops oxidizer dome and GG oxidizer purge.

13 Propellant bleed valves open allowing

14 propellant flow

Figure 6-4 (Sheet 2 of 2)

6-10

warms and raises the gas pressure to the relief valve setting. If this venting continues over a prolonged period, the total stored energy will be depleted. This limits the waiting period prior to a restart attempt to 6 hours.

PROPELLANTS

The propellants are LH_2 and lox. They are stored in the propellant tanks under controlled pressure to provide a minimum net positive suction head (NPSH) of 42 feet for the lox turbopump and 150 feet for the LH_2 turbopump during engine burn. Tank pressures are maintained by a system of tank vents and relief valves and a pressurizing system. The pressurizing system uses helium to pressurize the lox tank and helium or GH_2 to pressurize the LH_2 tank. The pressurizing and venting schedule is shown in figure 6-5.

OXYGEN/HYDROGEN BURNER

The oxygen/hydrogen (O_2/H_2) burner (figure 6-6) uses stage onboard propellants as an energy source to heat cold helium. The O_2/H_2 burner use spark plugs to ignite onboard propellants in its combustion chamber. The chamber is surrounded by three sets of coils. Tank propellants flow through one set of coils where they are vaporized and injected into the combustion chamber. The energy of combustion heats the cold helium in the other two sets of coils, causing it to vaporize. Gaseous helium flow is controlled by the LH_2/lox tank repressurization control module (dual valves). The burner produces 16 to 20 pounds thrust through the center of gravity as shown in figure 6-6.

PROPELLANT CONDITIONING

During filling operations, the prevalves are allowed to stay in the open position to provide a deadhead type chilldown of the feed system hardware (low pressure feed duct and engine pump), allowing temperature stabilization of this hardware prior to activation of the recirculation chilldown system. Approximately five minutes before liftoff, the prevalves are closed and recirculating chilldown flow is initiated (figure 6-7). It continues for about 14 minutes, until J-2 engine prestart. Since lox is already a subcooled liquid (no two-phase flow in the return line), prepressurization has negligible effect on the flowrate. The LH_2, however, becomes a subcooled liquid at prepressurization (eliminating two-phase flow in the return line), resulting in increased LH_2 flow.

Chilldown of the engine pumps, inlet ducting and the engine hardware for all engine starts is accomplished by separate lox and LH_2 chilldown systems. The purpose of the chilldown is to condition the ducting and engine to the proper temperature level, and to eliminate bubbles (two-phase flow) prior to prepressurization. The chilldown, along with the net positive suction head which is obtained by the proper pressure levels, provides the proper starting conditions.

Propellants from each tank are recirculated through the feed systems and return bleed lines by chilldown pumps. Check valves, prevalves, shutoff valves, and ducts control and route the fluids to perform the chilldown. Pneumatic pressure for operating the shutoff valves and prevalves is supplied by the stage pneumatic helium control bottle.

Figure 6-5

OXYGEN/HYDROGEN BURNER

CHAMBER PRESSURE PORT

CHAMBER TEMPERATURE PROBE PORT

INJECTOR NO. 2

IGNITER RECEPTACLE

COMBUSTION CHAMBER

ACCESS PORT

IGNITER RECEPTACLE

INJECTOR NO. 1

OXYGEN PRESSURE PORT

LH2 PRESSURE PORT

HELIUM OUTLET (LH2 TANK)

PRIMARY HELIUM COIL

OXYGEN TEMPERATURE PROBE PORT

LH2 INLET

LOX INLET

SECONDARY HELIUM OUTLET (LOX TANK)

GASEOUS OXYGEN OUTLETS

SECONDARY HELIUM OUTLET (LH2 TANK)

SHUTDOWN VALVE

SECONDARY HELIUM INLET (LH2 TANK)

HELIUM SECONDARY COILS

SECONDARY HELIUM INLET (LOX TANK)

NOZZLE

NOZZLE DIRECTION ADJUSTABLE ±5°

STA. 143.90 (NOZZLE EXIT)

S-IVB STAGE

10°

73"

III

26°30'

II

IV

I

STA. 100

J-? ENGINE

STA. 518.09 C. G. AT START OF SECOND BURN

Figure 6-6

PROPELLANT VENTING

The vent-relief subsystems (Figures 6-8 and 6-9) on the stage protect the propellant tanks against overpressurization and enable command venting at any time that controlled venting of tank pressure is required. The lox and LH$_2$ tanks each have a command relief and venting subsystem.

Both lox and LH$_2$ tank venting sequences provide for propulsive and non-propulsive venting (Figure 6-9). The lox tank is vented through its propulsive vent duct, located in the aft skirt, during loading. During flight, lox venting is normally a relief valve function and escaping vapors are routed through the non-propulsive vents. Specific command relief non-propulsive venting is accomplished by flight program or by ground command. During fill, LH$_2$ venting is routed through ground vent lines to the GH$_2$ burn pond. At T-40 seconds, the vent valves are commanded closed and the relief valves provide venting through the non-propulsive vents. The LH$_2$ tank is command vented through the

propulsive vents to provide propellant settling between the first and second burns. Specific venting sequences are noted in Figure 2-1 and Figure 6-5.

HELIUM SYSTEM

Helium is stored in nine cold helium spheres located in the LH$_2$ tank, eight ambient helium spheres mounted on the thrust structure and one sphere located inside the start bottle on the engine. The nine cold helium spheres supply cold helium for pressurization and repressurization of the lox tank and repressurization of the LH$_2$ tank. Five of the eight ambient helium spheres provide an alternate source of helium for repressurization of the LH$_2$ tank, two provide an alternate source of helium for repressurization of the lox tank and one provides pressure for operation of the stage pneumatic controls. The engine control helium sphere provides pressure for operation of the engine controls.

The ambient helium spheres are plumbed together as shown

Figure 6-7

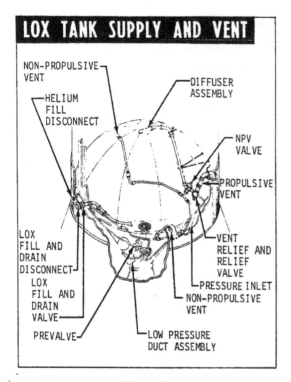

LOX TANK SUPPLY AND VENT

NON-PROPULSIVE VENT

HELIUM FILL DISCONNECT

DIFFUSER ASSEMBLY

NPV VALVE

PROPULSIVE VENT

LOX FILL AND DRAIN DISCONNECT

LOX FILL AND DRAIN VALVE

PREVALVE

VENT RELIEF AND RELIEF VALVE

PRESSURE INLET

NON-PROPULSIVE VENT

LOW PRESSURE DUCT ASSEMBLY

Figure 6-8

in Figure 6-9. This arrangement provides an in-flight reserve for the stage pneumatic controls helium sphere from the lox tank helium spheres and an in-flight reserve for the engine control helium sphere from the LH$_2$ tank helium spheres. The arrangement also permits venting of the two lox tank helium spheres, along with the five LH$_2$ tank helium spheres, through the LH$_2$ tank.

LOX SYSTEM

Lox is stored in the aft tank of the propellant tank structure (figure 6-8) at a temperature of -297 degrees F. Total volume of the tank is approximately 2830 cubic feet with an ullage volume of approximately 108 cubic feet. The tank is prepressurized between 38 and 41 psia and is maintained at that pressure during boost and engine operation. Gaseous helium is used as the pressurizing agent.

Lox Fill and Drain

The lox fill and drain valve is capable of allowing flow in either direction for fill or drain operations. During tank fill, the valve is capable of flowing 1000 gpm of lox at -297 degrees F at an inlet pressure of 51 psia. Pneumatic pressure for operating the fill and drain valve is supplied by the stage pneumatic control bottle.

Loading begins with a precooling flowrate of 500 gpm. When the 5% load level is reached, fast fill (1000 gpm) is initiated. At the 98% load level, fast fill stops and a slow fill at 300 gpm begins. A fast fill emergency cutoff sensor has been provided to compensate for a primary control cutoff failure. Slow fill is terminated at the 100% load level and this level is

then maintained by a replenish flowrate of 0 to 30 gpm, as required. The replenish flow is maintained through the complete lox tank prepressurization operation and the 100% lox load operation. Liquid level during fill is monitored by means of the lox mass probes.

Pressure sensing switches are used to control the tank pressure during fill. In the event of tank overpressurization (41 psia) the pressure switch sends a signal to close the lox ground fill valve.

Lox Engine Supply

A six-inch low pressure supply duct supplies lox from the tank to the engine. During engine burn, lox is supplied at a nominal flowrate of 430 pounds per second, and at a transfer pressure above 25 psia. The supply duct is equipped with bellows to provide compensating flexibility for engine gimbaling, manufacturing tolerances, and thermal movement of structural connections.

LOX TANK PRESSURIZATION

Lox tank pressurization is divided into three basic procedures. These procedures are called prepressurization, pressurization, and repressurization. The term prepressurization is used for that portion of the pressurization performed on the ground prior to liftoff. The term pressurization is used to indicate pressurization during engine burn periods, and lastly, repressurization indicates pressurization just before a burn period.

The pressurant used during the three lox tank pressurization procedures is gaseous helium. Cold helium from a ground source (1, figure 6-9) is used during the prepressurization period. This ground source of cold helium is also used to charge the nine cold helium storage spheres (2). The cold helium storage spheres, located in the LH$_2$ tank, supply cold helium for both the pressurization and repressurization periods. The ambient helium storage spheres (3), filled by ground support equipment (26), are the alternate source of helium for use during repressurization prior to the first and second burns.

The lox tank pressure is controlled by the flight control pressure switch (4) regardless of the pressurization procedure used. These switches control solenoid shutoff valves in each of the supply subsystems.

Prepressurization

At T-167 seconds, the lox tank is pressurized (figure 6-9) by ground support equipment. Pressure regulated cold helium (1) passes through the lox tank pressure control module (5) and flows (6) into the lox tank. When the lox tank pressure increases to 41 psia, the pressurization is completed and the flight control pressure switch shuts off the ground supply of cold helium.

Pressurization

After S-II/S-IVB separation at T$_4$ + 0.8, the lox tank flight pressure system is activated. When this system is activated, cold helium from cold helium storage spheres (2, figure 6-9) is routed through the lox tank pressure control module (5) to the J-2 engine heat exchanger (7). When the cold helium passes through the lox tank pressure control module (5), its

Figure 6-9

pressure is reduced to approximately 385 psia. As it passes through the heat exchanger it is expanded and routed to the lox tank. A small portion of cold helium bypasses the heat exchanger through a control orifice, and mixes with the hot gas prior to entering the lox tank. The flight control pressure switch (4) controls the operation of a solenoid valve in the lox tank pressure control module (5) to control lox tank pressure to 38-41 psia. The flow (8) of the helium to the lox tank is monitored by a backup pressure switch (9). If normal pressure regulation fails, this switch (9) will control lox tank pressure through operation of solenoid valves in the lox tank pressure control module (5). An S-IVB lox tank pressure reading becomes available in the command module (CM) at S-II/S-IVB separation. This pressure is sensed by the pressure transducer (12) and is relayed to the S-II FUEL/S-IVB OXID gauges (13) in the CM and, via telemetry, to the ground.

Repressurization

The normal repressurization procedure is initiated at T_6 + 48.3 seconds. It uses cold helium from the cold helium storage spheres (2, figure 6-9). The cold helium pressure is reduced to approximately 385 psia as it flows through the lox tank pressure control module (5). It next flows through the lox tank repressurization control (10), and into the O_2/H_2 burner. Should the regulator in the lox tank pressure control module (5) fail, the backup pressure switch (11) will maintain a pressure of 350-465 psia at the O_2/H_2 burner. The backup pressure switch controls the pressure by opening or closing valves in the lox tank repressurization control (10). As the cold helium is heated in the O_2/H_2 burner it expands and is routed to the lox tank. Pressure in the lox tank increases and is sensed by the flight control pressure switch (4) and the pressure transducer (12). The pressure switch (4) maintains lox tank pressure between 38-41 psia by opening and closing solenoid shutoff valves in the lox tank repressurization control (10). The pressure transducer (12) transmits a continuous pressure reading to telemetry and to the LV TANK PRESS gauges (13) in the CM. At T_6 + 496.6 seconds, cryogenic repressurization is switched off. Ambient repressurization is enabled at T_6 + 497.6 seconds and turned on at T_6 + 500.0 seconds. Ambient helium from the ambient helium storage spheres (3) flows through the lox tank, ambient helium, repressurization control module (14) to the lox tank. Here the pressure is sensed by the flight control pressure switch (4). The pressure switch (4) controls lox tank pressure by opening or closing the control valves in the lox tank, ambient helium, repressurization control module (14). Just before second burn, ambient repressurization is stopped (T_6 + 577.2 seconds).

NOTE

The ambient repressurization initiated prior to second burn is a back up procedure to ensure tank pressure in case of O_2/H_2 burner failure.

Lox Venting

The lox tank vent subsystem provides for controlled lox tank venting during normal stage operation and for pressure relief venting when tank overpressures occur. The lox tank venting subsystem includes a propulsive venting, vent and relief valve and a non-propulsive venting, latch open, non-propulsive vent valve. The vent and relief valve is pneumatically operated upon receipt of a ground command. Prior to loading, it is placed in the open position and the boiloff of lox during

loading is directed through the propulsive vent in the aft skirt of the stage (figure 6-8). When lox tank prepressurization commences, the valve is closed and placed in the relief position. The latch open non-propulsive vent valve is also operated by ground command. When it is commanded open, venting occurs through two non-propulsive vents placed 180 degrees apart (figure 6-8). When in the closed position, both valves act as relief valves. The non-propulsive vent valve is set to open at a maximum pressure of 43.5 psia. The vent and relief valve is set to open at a maximum of 45.5 psia and a minimum of 43.5 psia. This arrangement assures that relief venting will normally occur as a non-propulsive function. See figures 2-1 and 6-5 for sequence of operation

LH$_2$ SYSTEM

The LH$_2$ is stored in an insulated tank at less than -423 degrees F. Total volume of the tank is approximately 10,400 cubic feet with an ullage volume of approximately 300 cubic feet. The LH$_2$ tank is prepressurized to 28 psia minimum and 31 psia maximum.

LH$_2$ Low Pressure Fuel Duct

LH$_2$ from the tank is supplied to the J-2 engine turbopump through a vacuum jacketed low pressure 10-inch duct. This duct is capable of flowing 80-pounds per second at -423 degrees F and at a transfer pressure of 28 psia. The duct is located in the aft tank side wall above the common bulkhead joint. Bellows in this duct compensate for engine gimbaling, manufacturing tolerances, and thermal motion.

LH$_2$ Fill and Drain

Prior to loading, the LH$_2$ tank is purged with helium gas. At the initiation of loading, the ground controlled combination vent and relief valve is opened, and the directional control valve is positioned to route GH$_2$ overboard to the burn pond.

Loading begins with precool at a flow of 500 gpm. When the 5% load level is reached fast fill is initiated at a flow of 3000 gpm. At the 98% load level, fast fill stops and a slow fill at 500 gpm begins. A fast fill emergency cutoff sensor has been provided to compensate for a primary control cutoff failure. Slow fill is terminated at the 100% load level, and this level is then maintained by a replenish flowrate of 0 to 300 gpm, as required. The replenish flow is maintained through the LH$_2$ tank prepressurization operation.

Liquid level during fill is monitored by means of the LH$_2$ mass probes. A backup overfill sensor is provided to terminate flow in the event of a 100% load cutoff failure. Pressure sensing switches control the tank pressure during fill. In the event of tank over pressurization (31 psia) the pressure switch sends a signal to close the LH$_2$ ground fill valve.

LH$_2$ TANK PRESSURIZATION

LH$_2$ tank pressurization is divided into three basic procedures. These procedures are called prepressurization, pressurization, and repressurization. The term prepressurization is used for that portion of the pressurization performed on the ground prior to liftoff. The term pressurization is used to indicate pressurization during engine burn periods, and lastly, repressurization indicates pressurization just before the second burn period.

The pressurants used during the three LH$_2$ tank pressurization procedures are gaseous hydrogen (GH$_2$) and gaseous helium. Cold helium from a ground source (25, figure 6-9) is used during the prepressurization period. The cold helium storage spheres (2), located in the LH$_2$ tank, supply cold helium for use during the repressurization period. The five ambient helium storage spheres (15), filled by ground support equipment (26), supply an alternate source of helium for use during the repressurization period.

The LH$_2$ tank pressure is controlled by the flight control pressure switch (17) regardless of the pressurization procedure used. This switch controls solenoid shutoff valves in each of the supply subsystems.

Prepressurization

At T$_1$ -97 seconds, the LH$_2$ tank is prepressurized (figure 6-9) by ground support equipment. Cold helium (25) flows through the LH$_2$ tank pressurization module (18) and into the LH$_2$ tank. When the LH$_2$ tank pressure increases to 31 psia the flight control pressure switch (17) shuts off the ground supply of helium (25) to complete prepressurization.

Pressurization

Pressurization is controlled by the flight control pressure switch (17, figure 6-9) which opens or closes solenoid valves in the LH$_2$ tank pressurization module (18). Gaseous hydrogen (19) bled from the J-2 engine flows through the LH$_2$ tank pressurization module (18) to the LH$_2$ tank. As pressure in the LH$_2$ tank increases to 31 psia, the flight control pressure switch (17) closes valves in the LH$_2$ tank pressurization module to maintain tank pressure at 28-31 psia. This pressure is sensed by the pressure transducer (20) and is relayed to the S-IVB fuel gauges (21) in the CM and, via telemetry, to the ground. During pressurization the repressurization switch (16) operates in series with the flight control pressure switch to prevent operation of the cryogenic or ambient repressurization systems.

Repressurization

The normal repressurization procedure is initiated at T$_6$ + 48.1 seconds. It uses cold helium from the cold helium storage spheres (2, figure 6-9). The cold helium pressure is reduced to approximately 385 psia as it flows through the lox tank pressure control module (5). The cold helium next flows through the LH$_2$ tank repressurization control (22), and into the O$_2$/H$_2$ burner, it expands and is routed to the LH$_2$ tank. Pressure in the LH$_2$ tank increases and is sensed by the flight control pressure switch (17) and the pressure transducer (20). The pressure switch (17) maintains LH$_2$ tank pressure between 28-31 psia by opening and closing solenoid shutoff valves in the LH$_2$ tank repressurization control module (22). The pressure transducer (20) transmits a continuous pressure reading to telemetry and to the LV TANK PRESS gauges (21) in the CM. At T$_6$ + 496.7 seconds, cryogenic repressurization is switched off. Ambient repressurization is enabled at T$_6$ + 497.6 seconds and turned on at T$_6$ + 520 seconds. Ambient helium from the ambient helium storage spheres (15) flows through the LH$_2$ tank ambient helium repressurization control module (24) to the LH$_2$ tank. Here the pressure is sensed by the flight control pressure switch (17). The pressure switch (17) controls LH$_2$ tank pressure by opening or closing the control valves in the LH$_2$ tank ambient helium repressurization control module

(24). Just before second burn, ambient repressurization is stopped (T$_6$ + 577.4 seconds).

NOTE

The ambient repressurization initiated prior to second burn is a back up procedure to ensure tank pressure in case of O$_2$/H$_2$ burner failure.

LH$_2$ Venting

The LH$_2$ tank vent subsystem (Figure 6-9) is equipped to provide either propulsive or non-propulsive venting. Non-propulsive venting is the normal mode used.

The non-propulsive function is performed through the use of a ground controlled combination vent and relief valve which permits the option of routing the GH$_2$ through either the ground vent lines or through non-propulsive relief venting. The valve is in the ground vent line open position until T-40 seconds at which time it is positioned to the in-flight non-propulsive relief function. The non-propulsive vents are located 180 degrees apart to cancel the thrust.

The propulsive venting function is a command function which operates through two control valves upstream of the non-propulsive directional control valve. This mode vents the GH$_2$ through two propulsive vents located axial to the stage. Propulsive venting provides a small additional thrust, prior to the second J-2 engine burn, for propellant settling. Figure 2-1 and Figure 6-5 illustrate the sequential operation of the venting subsystems.

Directional Control Valve

The directional control valve is a two position valve provided to enable command routing of gaseous hydrogen (GH$_2$) through the ground vent line or through the flight nonpropulsive vents. During LH$_2$ loading, the valve is positioned to vent GH$_2$ through the ground line to the hydrogen burn pond. At T-40 seconds the directional control valve is positioned for flight venting of GH$_2$ through the nonpropulsive vents.

PROPELLANT DUMP

A propellant dump is initiated in T$_8$ once the S-IVB/IU has completed an evasive maneuver to safely clear the CSM/LM following separation. This sequence dumps the lox remaining in the lox tank through the engine, thereby applying a delta retro-grade velocity to the S-IVB/IU which places it in solar orbit. As a part of the propellant dump the tanks and helium bottles are vented as shown in figure 6-5 to passivate the stage.

PNEUMATIC CONTROL

The pneumatic control system (figure 6-10) provides pressure for all pneumatically operated valves on the stage and for the engine start tank vent valve on the J-2 engine. The pneumatic control system is filled with gaseous helium from ground support equipment to 3200 ± 100 psig. The onboard pneumatic control system consists of the helium fill module, an ambient helium sphere and a pneumatic power control module. Backup to the control system helium sphere is provided by connection through check valves with the two lox system ambient helium spheres.

PNEUMATIC CONTROL SUBSYSTEM

Figure 6-10

The helium fill module regulates and reduces the incoming supply to 490 ± 25 psia for operation of control valves during preflight activities. The ambient helium storage spheres are initially pressurized to 950 psia, and are capable of supplying operating pressure to stage control valves at that pressure. After propellant loading has begun, and the cold helium bottles are chilled down, the pressure is raised to 3100 psia and both the ambient and cold helium bottles are then completely pressurized to their flight pressure of 3200 psia by the time the LH_2 tank reaches a 92% percent load level.

The pneumatic power control module is set at 475 psig which is equivalent to 490 psia on the ground and 475 psia in orbit. These pressure levels are essential to the operation of the LH_2 directional control valve, the propulsion vent shutoff valve, the lox and LH_2 fill and drain valves, the lox and LH_2

turbopump turbine purge module, the lox chilldown pump purge control, the lox and LH_2 prevalves and chilldown and shutoff valves, the lox tank vent/relief valves, the LH_2 propulsive vent valve, and the J-2 engine GH_2 start system vent/relief valve. Each pneumatically operated component is attached to a separate actuation control module containing dual solenoids, which provide on-off control.

The pneumatic control system is protected from over-pressure by a normally open, pressure switch controlled, solenoid valve. This switch maintains system pressure between 490 - 600 psia.

FLIGHT CONTROL

The flight control system uses two systems for flight and

attitude control. During powered flight, steering is accomplished by gimbaling the J-2 engine for pitch and yaw control, and by operating the APS engines for roll control. Steering during coast flight is by use of the APS engines alone.

ENGINE GIMBALING

During the boost and separation phase, the J-2 engine is commanded to the null position to prevent damage by shifting. The engine is also nulled before engine restart to minimize the possibility of contact between the engine bell and the interstage at S-II/S-IVB separation, and to minimize inertial effects at ignition. The engine is gimbaled (figure 6-11) in a 7.0 degrees square pattern by a closed loop hydraulic system. Mechanical feedback from the actuator to the servovalve completes the closed engine position loop.

When a steering command is received from the flight control computer, a torque motor in the servovalve shifts a control flapper to direct the fluid flow through one of two nozzles, the direction being dependent upon signal polarity.

Two actuators are used to translate the steering signals into vector forces to position the engine. The deflection rates are proportional to the pitch and yaw steering signals from the flight control computer.

HYDRAULIC SYSTEM

Major components of the hydraulic system (figure 6-12) are an engine driven hydraulic pump, an electrically driven auxiliary hydraulic pump, two hydraulic actuator assemblies, and an accumulator/reservoir assembly.

Hydraulic Pumps

The engine driven hydraulic pump is a variable displacement type driven directly from the engine oxidizer turbopump. In normal operation, the pump delivers up to 8 gpm under continuous working pressure.

The auxiliary hydraulic pump is an electrically driven pump which is capable of supplying a minimum of 1.5 gpm of fluid to the system. This pump supplies pressure for preflight checkout, to lock the J-2 engine in the null position during boost and separation phase, and as emergency backup. During orbit, the auxiliary pump, controlled by a thermal switch, circulates the hydraulic fluid to maintain it between +10 degrees F and +40 degrees F. The auxiliary pump is enabled before liftoff and during coast periods.

Accumulator/Reservoir Assembly

The accumulator/reservoir assembly (figure 6-12) is an integral unit mounted on the thrust structure. The reservoir section is a storage area for hydraulic fluid and has a maximum volume of 167 cubic inches.

During system operation, between 60 and 170 psig is maintained in the reservoir (figure 6-12) by two pressure operated pistons contained in the accumulator section. In addition to maintaining pressure in the reservoir, the system accumulator supplies peak system demands and dampens high pressure surging.

Hydraulic Actuators - Pitch and Yaw

The pitch and yaw actuators and servovalve (figure 6-12) are integrally mounted and are interchangeable. The actuators are linear and double acting. During powered flight, pitch and yaw control is provided by gimbaling the main engine, the two actuator assemblies providing deflection rates proportional to pitch and yaw steering signals from the flight control computer.

AUXILIARY PROPULSION SYSTEM

The S-IVB auxiliary propulsion system provides three axis stage attitude control (figure 6-13) and main stage propellant control during coast flight.

APS CONSTRUCTION

The APS engines are located in two modules, 180 degrees apart on the aft skirt of the S-IVB stage (see figure 6-14). The modules are detachable and are easily checked or replaced. Each module contains four engines: three 150-pound thrust control engines (figure 6-15) and one 70-pound thrust ullage engine (figure 6-16). Each module contains its own oxidizer, fuel, and pressurization systems. A positive expulsion, propellant feed subsystem is used to ensure that hypergolic propellants are supplied to the engine under zero or random gravity conditions. This subsystem consists of separate fuel and oxidizer propellant tank assemblies (figure 6-17), each containing a bladder for propellant expulsion, individual propellant control modules that filter the propellants and provide auxiliary ports for subsystem servicing operations, and a propellant manifold for distribution of propellants to the engines.

Nitrogen tetroxide (N_2O_4), is used as an oxidizer and monomethyl hydrazine (MMH), is used as fuel for these engines. The 150-pound thrust engines utilize eight control valves (four for fuel, four for oxidizer) in a fail safe, series parallel arrangement. The 70-pound ullage engine uses only single valves on both the fuel and oxidizer lines.

APS OPERATION

Two of the three 150-pound thrust engines in each module control roll and yaw, while one controls pitch. The 70-pound thrust engine in each module is used to settle the propellants in the propellant tanks. Just prior to J-2 engine restart, propellant control is accomplished by firing the 70-pound thrust engines. The attitude control engines are fired for attitude correction. The minimum engine firing pulse duration is approximately 70 milliseconds. Typical operation (figure 6-18) involves a signal from the Instrument Unit (IU) to energize the pitch control relays which open the eight normally-closed, quad-redundant propellant valves (4 oxidizer, 4 fuel) arranged in two series parallel circuits. Valve assembly failure cannot occur unless two valves fail open in series (propellant drain), or two valves fail closed in parallel (oxidizer or fuel starvation). Single valve failures do not affect engine efficiency or performance. The attitude deviation dead band for roll, pitch, and yaw is ± 1 degree.

The attitude engine control signal is composed of an attitude error signal and a vehicle turning rate signal. The body mounted control EDS rate gyros, located in the IU, supply the vehicle turning rate information. Attitude error information may come from two sources. When the IU is

THRUST VECTOR CONTROL

ACCUMULATOR/RESERVOIR

SERVOACTUATOR (PITCH)

SERVOACTUATOR (YAW)

ENGINE DRIVEN
HYDRAULIC PUMP

AUXILIARY
HYDRAULIC PUMP

LEGEND

PRESSURE
SUPPLY
RETURN

Figure 6-11

controlling the vehicle attitude, error information is supplied by the LVDC through the flight control computer. When the spacecraft is controlling the vehicle, attitude error information is obtained from the spacecraft. Limits are imposed by the LVDC for IU generated attitude error signals, and by the IU flight control computer for spacecraft generated attitude error signals. It should be noted that attitude error limits of ± 2.5 degrees in pitch and yaw and ±3.5 degrees in roll are imposed on the attitude error signals. These error signals are used only by the auxiliary propulsion system. The attitude error signals from the spacecraft may originate in the Apollo navigation, guidance and control system or may be generated by the astronaut through manual control. In any case, the limiters in the IU flight control computer will limit the angular rate. These limits prevent excessive propellant usage which would result from large angular rate commands, while driving the vehicle to the desired attitude.

The Apollo spacecraft attitude reference system can follow the instantaneous vehicle attitude. This is accomplished by driving the command display unit servomotor with an error signal which is formed by differencing the commanded and actual gimbal angles. When the astronaut wishes to maintain a particular attitude orientation, he can use the computer to set the command display unit command resolver to the desired gimbal value. The difference between the commanded and actual gimbal angles results in an error signal which is resolved into vehicle coordinates and is given to the IU flight control computer as an attitude error signal. The S-IVB attitude control system then operates in the limit cycle mode about this command attitude.

ELECTRICAL

The electrical system of the S-IVB stage is comprised of two major subsystems: the electrical power subsystem which consists of all power sources on the stage; and the electrical control subsystem which distributes power and control signals to various loads throughout the stage.

BATTERIES

On board power is supplied by four zinc silver-oxide batteries. Two are located in the forward equipment area and two in the aft equipment area. These batteries are activated and installed in the stage during the final prelaunch preparations. Heaters and instrumentation probes are an integral part of each battery. See figure 6-19 for a table of battery characteristics.

POWER DISTRIBUTION

Two main forward busses and two main aft busses supply electrical power to all stage systems. Busses are electrically isolated from each other with each main bus utilizing a power transfer switch to switch from GSE power to stage mounted batteries. the power is distributed as shown in figure 6-20.

CHILLDOWN INVERTERS

The chilldown inverters are three-phase, 1500 volt-amp, solid state power conversion devices. The purpose of the chilldown inverters is to provide electrical power to the motor driven pumps for circulation of the lox and LH_2, to ensure propellant temperature stabilization at the J-2 engine inlets.

HYDRAULIC SYSTEM COMPONENTS

Figure 6-12

ATTITUDE CONTROL

Figure 6-13

PROPELLANT UTILIZATION (PU) STATIC INVERTER-CONVERTER

The static inverter-converter is a solid state power supply which provides all the regulated ac and dc voltages necessary to operate the S-IVB stage propellant utilization electronics assembly.

The static inverter-converter converts a 28 vdc input into the following outputs:

1. 115 vac 400 Hz required to drive propellant utilization valve positioning motors and bridge rebalancing servomotors,

2. 2.5 vac, (peak-to-peak), square wave required to convert propellant utilization error signal to alternating current,

3. 5 vdc required to provide excitation for propellant utilization fine and coarse mass potentiometers,

Figure 6-14

4. regulated 21 vdc required for propellant utilization bridges,

5. 117.5 vdc floating supply used to provide propellant utilization summing potentiometer excitation,

6. 49 vdc floating supply required to provide excitation for valve feedback potentiometers.

EXCITATION MODULES

The 5-volt and 20-volt excitation modules are transistorized power conversion devices which are used to convert the 28-vdc to the various regulated voltages required by the instrumentation, signal conditioning, and emergency detection system transducers.

ELECTRICAL CONTROL SUBSYSTEM

The electrical control subsystem function is to distribute the command signals required to control the electrical components of the stage. The major components of the electrical control subsystem are the power and control distributors, the sequencer assemblies, and the pressure sensing and control devices.

SEQUENCER ASSEMBLIES

The two major elements in the stage sequencing system are the switch selector and the stage sequencer. During flight, sequencing commands are received from the IU. Each command is in digital form and consists of an 8-bit word accompanied by a "read" pulse. The commands are interpreted by the S-IVB stage at the switch selector.

Switch Selector

The switch selector is an electronic assembly utilized as the primary device for controlling the inflight sequencing of the stage. A switch selector is utilized in each stage of the launch vehicle.

The switch selector consists of relays, a diode matrix, and low-power transistor switches used as relay drivers controlled by binary-coded signals from the LVDC in the IU. The function of the switch selector is to operate magnetically latching relays in the sequencer and power distribution assemblies.

The switch selector provides electrical isolation between the IU and the S-IVB stage systems and decodes digital information from the IU into discrete output commands. Capability exists to provide 112 discrete commands to the stage sequencer.

Stage Sequencer

The stage sequencer operates upon receipt of discrete inputs from the switch selector (and other S-IVB stage subsystems) and initiates S-IVB flight functions by supplying or removing power from the appropriate equipment. Sequence circuits perform logistical gating of inputs necessary for sequencing control with as few timed commands from the IU as possible. It controls only those functions established as sequencing events.

150-POUND THRUST APS ENGINE

ABLATIVE
EXIT CONE
LINER

COATED MOLY
THROAT INSERT

5.750 INCHES
CERAMIC
ABLATIVE
LINER

QUAD
REDUNDANT
VALVE
ASSEMBLY

HEAT SINK
INJECTOR

FIBERGLASS
FLAT WRAP

RADIATION SHIELD (LOWER)

RADIATION SHIELD (UPPER)

Figure 6-15

70-POUND THRUST APS ENGINE

PROPELLANT VALVES

ELECTRICAL CONNECTOR

MOUNTING CAN

GLASS WRAP

ASBESTOS WRAP

PARALLEL WRAP ABLATIVE

INJECTOR

CERAMIC LINER

PARALLEL WRAP ABLATIVE

90° ORIENTED ABLATIVE

CERAMIC INSERT

Figure 6-16

Figure 6-17

PRESSURE SWITCHES

Calibratable pressure switches (calips) are used on the S-IVB stage to perform various control functions. For example:

1. LH$_2$ Tank System

   a. Pressurization, ground fill, valve control,

   b. LH$_2$ tank pressure control backup,

2. Lox Tank System

   a. Ground fill, valve control, pressurization,

   b. Lox tank pressure control backup,

3. Pneumatic Power System

   a. Regulator backup,

   b. Engine purge,

   c. Lox chilldown pump container purge,

These pressure switches are located in either the aft section (thrust structure) or the forward section (interstage), depending upon the pressure control system in which the switch is used.

The calips pressure switch employs two pressure ports, each isolated from the other. The test port provides for remote checkout without disconnecting or contaminating the primary pressure system. The test pressure settings are calibrated during manufacture of the switch to provide an accurate indication of the system pressure settings.

Calips pressure switches utilize a single Belleville spring which provides "snap" response to actuation or deactuation pressures. This response provides switching of 28 vdc power to relays in the stage sequencer for operation and control of propulsion system solenoid valves.

PROPELLANT MANAGEMENT

The propellant management systems illustrated in figure 6-21 provide a means of monitoring and controlling propellants during all phases of stage operation. Continuous capacitance probes and point level sensors in both the LH$_2$ and lox tanks monitor propellant mass. Point level sensors are used during the propellant loading sequence to indicate to the GSE the level of propellants in the tanks. Level sensors signal the LVDC during flight so that engine cutoff can be accomplished smoothly at propellant depletion. The capacitance probes provide outputs which are used to operate a propellant utilization (PU) control valve. The PU valve is a rotary type valve which controls the quantity of lox flowing to the engine.

PROPELLANT UTILIZATION SUBSYSTEM

Prior to S-IVB ignition for first burn, the PU valve is commanded to its null position to obtain an engine mixture ratio (MR) of approximately 5:1 for start and remains at this value for the remainder of the burn. Prior to the second start the PU valve is commanded to its full open position to obtain

ATTITUDE AND ULLAGE ENGINE CONTROL

Figure 6-18

a MR of approximately 4.5:1 during the start sequence. Approximately 100 seconds after 90% thrust is attained in the first opportunity mission the PU valve is commanded back to its null position for a MR of approximately 5:1 where it remains for the rest of the burn. In the second opportunity mission the PU valve is commanded back to its null position at approximately 2.5 seconds after 90% thrust is attained. Operation of the engine in the 4.5:1 MR results in a proportionally larger consumption of fuel than operation in the 5:1 MR. The longer operation in the 4.5:1 MR during the first opportunity mission consumes the fuel which is carried to compensate for boiloff during orbit if the second opportunity mission is flown.

The excursion effect caused by varying the MR is illustrated in figure 5-8.

INSTRUMENTATION AND TELEMETRY

The S-IVB stage instrumentation and telemetry system monitors functional operation of stage systems and transmits these data to GSE. Prior to liftoff, telemetry data is transmitted over coaxial cables to the GSE. During flight RF antennae convey the data to ground stations. See figure 6-22 for a block diagram of the instrumentation and telemetry system.

MEASUREMENT SYSTEM

The measurement system monitors stage systems parameters and discretes. These data signals are conditioned to make them compatible and are then fed to the telemetry system for transmission to the ground.

BATTERY CHARACTERISTICS

	CHARACTERISTICS			
TYPE	Dry Charge			
MATERIAL	Zinc-silver oxide			
ELECTROLYTE	Potassium Hydroxide (KOH) in pure water.			
	FORWARD NO. 1	FORWARD NO. 2	AFT NO. 1	AFT NO. 2
CURRENT RATING	228 AH	25 AH	228 AH	67 AH
NOMINAL VOLTAGE	28 vdc (2 UNITS)	28 vdc	28 vdc (2 UNITS)	56 vdc (2 UNITS IN SERIES)

Figure 6-19

Numerous measuring devices (transducers), of a wide variety, are used to monitor stage systems. Conditions monitored include such things as temperature, pressure, flowrate, position, discretes, liquid levels, voltages, current, frequency and RPM.

TELEMETRY SYSTEM

The telemetry system accepts the signals produced by the measuring portion of the instrumentation system and transmits them to the ground stations. On vehicles SA-507 and SA-508 both a SS/FM link (S1) and a PCM/DDAS link (P1) are used. On vehicle SA-509 only the PCM/DDAS link (P1) is used.

The SS/FM link includes a single sideband telemetry assembly with fifteen input data channels which are frequency-division multiplexed over a spectrum of 70 KHz. Each channel has a frequency response suitable for vibration and acoustic type measurements. An isolation amplifier accepts the signal from the single sideband assembly and provides two isolated outputs. One output is supplied to a Model II RF assembly and the other output is used for ground checkout.

The PCM/DDAS link incorporates a Model 301 PCM/DDAS assembly which accepts both analog and digital data. The analog data is received from a MUX-270 located in the S-IVB and a MUX-270 located in the IU. The digital data is received from digital transducers and from the Model ID remote digital sub-multiplexer assembly (RDSM). The PCM/DDAS assembly converts the analog data into binary code and arranges the data along with the digital input into the desired format for transmittal by a Model II RF assembly.

In addition, there is one MUX-270 on the S-IVB stage which supplies analog data to the PCM/DDAS link (Pl) in the Instrument Unit. This MUX-270 receives inputs directly from transducers and from a low-level remote analog submultiplexer, Model 102 (RASM-102).

The Model II RF assembly provides a carrier which is frequency-modulated by the composite output signal of the single sideband assembly or the PCM/DDAS assembly. The transmitter provides a 20-watt nominal output to the vehicle antenna system.

The antenna system includes RF power dividers, switches, couplers and detectors. The RF detectors monitor transmitter power and reflected power to check the RF transmission efficiency. The antennae are two folded dipoles located on opposite sides of the forward skirt. They have a radiation pattern such that two ground stations can see the stage simultaneously.

ORDNANCE

The S-IVB ordnance systems include the separation, ullage rocket, ullage rocket jettison, and propellant dispersion (flight termination) systems.

SEPARATION SYSTEM

The third plane separation system for S-II/S-IVB is located at the top of the S-II/S-IVB interstage (figure 5-20). The separation plane is at station 2746.5. Ordnance for the third plane separation consists of two exploding bridgewire (EBW) firing units, two EBW detonators, one detonator block assembly and a detonating fuse assembly (figure 6-23). The EBW firing units are on the S-II/S-IVB interstage slightly below the third separation plane. The leads of the EBW firing units are attached to the EBW detonators which are installed in the detonator block assembly. The detonator block assembly is mounted just inside the skin of the vehicle, and the ends of the detonating fuse assembly are installed within the detonator block assembly. The detonating fuse assembly is mounted around the periphery of the vehicle beneath the tension strap.

ELECTRICAL POWER DISTRIBUTION

Figure 6-20

The two EBW firing units for third plane separation provide redundant signal paths for initiation of the detonating fuse assembly. The function of the ordnance train is similar to that described in the separation system discussion in Section V. Detonation of the detonating fuse assembly severs the tension strap attaching the S-II/S-IVB interstage at station 2746.5 (figure 5-19).

A sequence of events for S-IC/S-II/S-IVB separations and a block diagram of the separation systems is contained in figure 5-20.

At the time of separation, four retrorocket motors mounted on the interstage structure below the separation plane fire to decelerate the S-II stage. For information on the S-II retrorocket system, refer to Section V.

ULLAGE ROCKET SYSTEM

To provide propellant settling and thus ensure stable flow of lox and LH$_2$ during J-2 engine start, the S-IVB stage requires a small acceleration. This acceleration is provided by two ullage rockets.

The S-IVB ullage rocket system (figure 6-23) consists of two EBW firing units, two EBW detonators, two confined detonating fuse (CDF) manifolds, nine CDF assemblies, two separation blocks, four CDF initiators, and two ullage rockets. The EBW firing units, EBW detonators, and CDF manifolds are mounted on the S-IVB aft skirt. The CDF assemblies connect the manifolds to the separation blocks and then to the CDF initiators. The rockets are within fairings mounted diametrically opposite each other on the S-IVB aft skirt. The rockets are canted outward from the vehicle to reduce effects of exhaust impingement, and to reduce the resulting moment if one rocket fails.

A separation block is used between the stage and each ullage rocket to allow jettison and maintain CDF continuity. The separation block, an inert item, is located on the skin of the S-IVB aft skirt under the ullage rocket fairing. Each block consists of two machined pieces of aluminum. The upper piece holds the ends of the CDF assemblies to the initiators, while the lower piece holds the CDF assemblies from the manifolds. The separation block forms a housing or connector that holds the CDF assembly ends together to

Figure 6-21

INSTRUMENTATION AND TELEMETRY SYSTEM

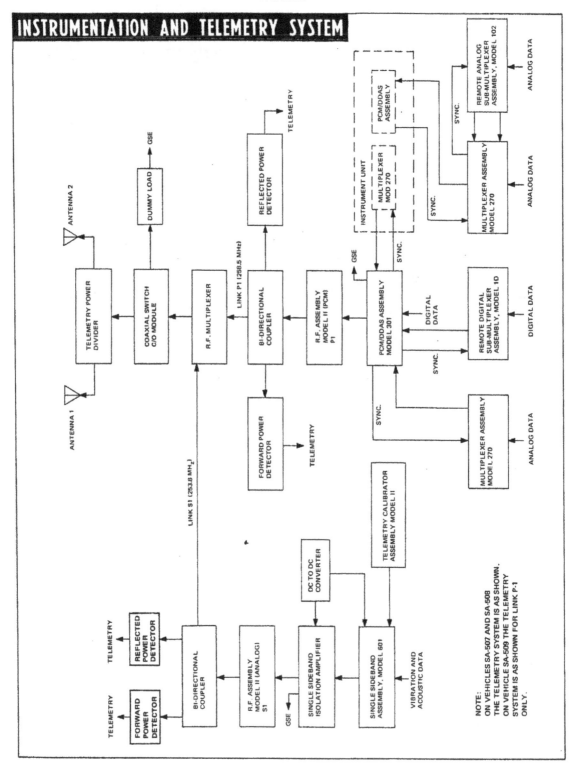

Figure 6-22

ensure propagation and to contain the detonation of the connection. At jettison, the block slips apart with the lower portion remaining on the stage and the upper portion falling away with the rocket and fairing.

Each ullage rocket has a single grain, five point configuration, internal burning, polymerized solid propellant that is case bonded in a 4135 steel case. The propellant weighs approximately 58.8 pounds, and burns for 3.87 seconds, developing a thrust of 3390 pounds (175,000 feet, 70 degrees F).

The firing sequence begins with the arming of the EBW firing units by charging the storage capacitors to 2300 volts. At S-II engine shutdown, the EBW units receive a trigger signal which discharges the storage capacitors, releasing high energy pulses to the EBW detonators, and thereby exploding the bridgewires. The resulting detonations propagate through the CDF manifolds, CDF assemblies, separation blocks and to the

termination of vehicle flight during the S-IVB first engine firing boost burn period if the vehicle flight path varies beyond its prescribed limits or if continuation of vehicle flight creates a safety hazard. The S-IVB PDS may be safed after the launch escape tower is jettisoned. The system is installed in compliance with Air Force Eastern Test Range (AFETR) Regulation 127-9 and AFETR Safety Manual 127-1.

The S-IVB PDS is a dual channel, parallel redundant system composed of two segments (figure 6-23). The radio frequency segment receives, decodes and controls the propellant dispersion commands. The ordnance train segment consists of two EBW firing units, two EBW detonators, one safety and arming (S&A) device (shared by both channels), seven CDF assemblies, two CDF tees, and three linear shaped charge (LSC) assemblies.

Should emergency termination become necessary, two coded messages are transmitted to the launch vehicle by the range

Figure 6-23 (Sheet 1 of 2)

MAJOR ORDNANCE COMPONENTS

ULLAGE ROCKET SYSTEM

ULLAGE ROCKET JETTISON SYSTEM

PROPELLANT DISPERSION SYSTEM

THIRD PLANE SEPARATION SYSTEM

Figure 6-23 (Sheet 2 of 2)

CDF initiators which cause the ullage rockets to ignite. A crossover CDF assembly between CDF manifolds provides redundancy and added system reliability.

ULLAGE ROCKET JETTISON SYSTEM

To reduce weight, the ullage rockets and their fairings are jettisoned after J-2 engine start. The system, located on the S-IVB aft skirt, uses two EBW firing units, two EBW detonators, one detonator block, two CDF assemblies, four frangible nuts, and two spring-loaded jettison assemblies (figure 6-23).

The EBW firing units are armed by charging their storage capacitors to 2300 volts about five seconds after the S-IVB ullage rockets have stopped firing. A trigger signal releases the high voltage pulse to explode the bridgewire in the EBW detonator. Either detonator will detonate both CDF assemblies (figure 6-23) through the detonator block. The detonation propagates through the CDF assemblies to detonate and fracture the frangible nuts. This frees the bolts that secure the ullage rocket and fairing assemblies to the aft skirt. The spring loaded jettison assemblies propel the spent rocket and fairing assemblies away from the vehicle.

PROPELLANT DISPERSION SYSTEM

The S-IVB propellant dispersion system (PDS) provides for safety officer. The first command arms the EBW firing units and initiates S-IVB stage engine cutoff. The second command, which is delayed to permit charging of the EBW firing units, discharges the storage capacitors across the exploding bridgewires in the EBW detonators mounted on the S&A device. The resulting explosive wave propagates through the S&A device inserts and through the remainder of the ordnance train to sever the LH_2 and lox tanks.

A description of the S&A device is included in the PDS discussion in Section IV.

The linear shaped charges for the LH_2 and lox tanks are RDX loaded at 150 grains per foot. Two assemblies are used to cut two 20.2-foot long parallel openings in the side of the LH_2 tank. One assembly is used to cut a 47-inch diameter hole in the bottom of the lox tank.

Following S-IVB engine cutoff at orbit insertion, the PDS is electrically safed by ground command.

MAJOR DIFFERENCES

▮▶ *The telemetry system for the S-IVB stages of vehicles SA-506 and SA-509 consist of one PCM/DDAS link while the telemetry system of the S-IVB stages of vehicles SA-507 and SA-508 includes one SS/FM link and one PCM/DDAS link.*

This Space Intentionally Left Blank

SECTION VII

INSTRUMENT UNIT

TABLE OF CONTENTS

INTRODUCTION

The Instrument Unit (IU) is a cylindrical structure installed on top of the S-IVB stage (see figure 7-l). The IU contains the guidance, navigation, and control equipment which will guide the vehicle through its earth orbits and subsequently into its mission trajectory. In addition, it contains telemetry, communications, tracking, and crew safety systems, along with their supporting electrical power and environmental control systems.

This section of the Flight Manual contains a description of the physical characteristics and functional operation for the equipment installed in the IU.

SATURN INSTRUMENT UNIT

Figure 7-1

STRUCTURE

The basic IU structure is a short cylinder fabricated of an aluminum alloy honeycomb sandwich material (see figure 7-2). The structure is fabricated from three honeycomb sandwich segments of equal length. The top and bottom edges are made from extruded aluminum channels bonded to the honeycomb sandwich. This type of construction was selected for its high strength-to-weight ratio, acoustical insulation, and thermal conductivity properties. The cylinder is manufactured in three 120 degree segments (figure 7-4), which are joined by splice plates into an integral structure. The three segments are the access door segment, the flight control computer segment, and the ST-124-M3 segment. The access door segment has an umbilical door, as well as an equipment/personnel access door. The access door has the requirement to carry flight loads, and still be removable at any time prior to flight.

Attached to the inner surface of the cylinder are cold plates which serve both as mounting structure and thermal conditioning units for the electrical/electronic equipment. Mounting the electrical/electronic equipment around the inner circumference of the IU leaves the center of the unit open to accommodate the convex upper tank bulkhead of the S-IVB stage and the landing gear of the Lunar Module (LM).

Cross section "A" of figure 7-3 shows equipment mounting pads bolted and bonded to the honeycomb structure. This method is used when equipment is not mounted on thermal conditioning cold plates. The bolts are inserted through the honeycomb core, and the bolt ends and nuts protrude through the outside surface. Cross section "B" shows a thermal conditioning cold plate mounting panel bolted to brackets which, in turn, are bolted on the honeycomb structure. The bolts go through the honeycomb core with the bolt heads protruding through the outer surface. Cross section "C" shows the cable tray supports bolted to inserts, which are potted in the honeycomb core at the upper and lower edges of the structure.

Figure 7-2

Figure 7-4 shows the relative locations of all equipment installed in the IU.

ENVIRONMENTAL CONTROL SYSTEM

The environmental control system (ECS) maintains an acceptable operating environment for the IU equipment during preflight and flight operations. The ECS is composed of the following:

1. The thermal conditioning system (TCS) which maintains a circulating coolant temperature to the electronic equipment of 59 ($\pm$ 1) degrees F.

2. The preflight purging system which maintains a supply of temperature and pressure regulated air/GN$_2$ in the IU/S-IVB equipment area.

3. The gas-bearing supply system which furnishes GN$_2$ to the ST-124-M3 inertial platform gas bearings.

4. The hazardous gas detection sampling equipment which monitors the IU/S-IVB forward interstage area for the presence of hazardous vapors.

THERMAL CONDITIONING SYSTEM

Thermal conditioning panels, also called cold plates, are located in both the IU and S-IVB stage; there are up to sixteen in each. Each cold plate contains tapped bolt holes in a grid pattern which provides flexibility of component mounting.

The cooling fluid circulated through the TCS is a mixture of 60% methanol and 40% demineralized water, by weight. Each cold plate is capable of dissipating at least 420 watts.

A functional flow diagram is shown in figure 7-5. The main coolant loop is the methanol-water mixture. Two heat exchangers are employed in the system. One is used during the preflight mode and employs a GSE supplied, circulating, methanol-water solution as the heat exchanging medium. The other is the flight mode unit, and it employs demineralized water and the principle of sublimation to effect the cooling.

The manifold, plumbing, and both accumulators are manually filled during the prelaunch preparations. The accumulators serve as dampers to absorb pressure fluctuations and thermal expansion, and also as reservoirs to replace lost or expended fluids. There are flexible diaphragms in each accumulator, backed by low pressure GN$_2$.

During operation of the TCS, the methanol-water coolant is circulated through a closed loop by electrically driven, redundant pumps. The flow is from the heat exchanger past the accumulator, through the pumps, through a temperature sensor to an orifice assembly. The orifice assembly diverts part of the coolant to the cold plates in the S-IVB stage and part to the cold plates, gas bearing heat exchanger, inertial platform, LVDC/LVDA and flight control computer in the IU. Return flow is through a modulating flow control assembly which regulates the amount of coolant flowing into the heat exchangers or around them. Operation of this valve is based on fluid temperature.

During the preflight mode, the sublimator heat exchanger is inactive due to the high ambient pressure (one atmosphere),

INSTRUMENT UNIT STRUCTURAL DETAILS

EQUIPMENT MOUNTING PADS

COLD PLATE

CABLE TRAY SUPPORTS

A

B

C

SLA INTERFACE

HORIZONTAL SPLICE PLATE

INNER SPLICE PLATE

OUTER SPLICE PLATE

TYPICAL SPLICE PLATE ATTACHMENT

0.200 IN.

36 IN.

0.125 IN.

UPPER MOUNTING RING

0.95 IN. FULL WALL THICKNESS

HIGH DENSITY CORE

LOW DENSITY CORE

0.030 ALUM 7075-T6 OUTER SKIN

0.020 ALUM ALLOY 7075-T6 INNER SKIN

LOWER MOUNTING RING

S-IVB INTERFACE TYPICAL SECTIONAL VIEW

1.25 IN.

Figure 7-3

and a solenoid valve which blocks the water flow. The preflight heat exchanger is operated from the GSE, and cools the closed loop fluid.

Approximately 180 seconds after liftoff, the water solenoid valve is opened and the sublimator heat exchanger becomes active. During the period between GSE disconnect and sublimator activation, the residual cooling in the system is sufficient to preclude equipment overheating.

The sublimator element is a porous plate. Since the sublimator is not activated until approximately 180 seconds after launch, the ambient temperature and pressure outside the porous plates are quite low. Water flows readily into the porous plates and attempts to flow through the pores. However, the water freezes when it meets the low temperature of the space environment, and the resulting ice blocks the pores (see figure 7-6).

As heat is generated by equipment, the temperature in the methanol-water solution rises. This heat is transferred within the sublimator to the demineralized water. As the water temperature rises, it causes the ice in the pores to sublime. The vapor is vented overboard. As the heat flow decreases, ice plugs are formed in the pores, decreasing the water flow. Thus, the sublimator is a self-regulating system. GN_2 for the methanol-water and water accumulators is stored in a 165 cubic inch sphere in the IU, at a pressure of 3,000 psig. The sphere is filled by applying high pressure GN_2 through the umbilical. A solenoid valve controls the flow into the sphere, and a pressure transducer indicates to the GSE when the sphere is pressurized. The output of the sphere is applied to the accumulators through a filter and a pressure regulator, which reduces the 3,000 psig to 15 psia. An orifice regulator

further reduces the pressure at the accumulator to 5 psia, the differential of 10 psia being vented into the IU.

PREFLIGHT AIR/GN_2 PURGE SYSTEM

The preflight air/GN_2 purge system (see figure 7-7) consists primarily of flexible ducting located above the IU payload interface. The system distributes ground supplied, temperature and pressure regulated, filtered air or GN_2 through openings in the ducting. During preflight phases, ventilating air is furnished. During fueling, inert GN_2 is furnished to prevent accumulation of a hazardous and corrosive atmosphere.

GAS BEARING SUPPLY

Gaseous nitrogen, for the ST-124-M3 stable platform, is stored in a two cubic foot sphere in the IU, at a pressure of 3,000 psig (see figure 7-5). The sphere is filled by applying high pressure GN_2 through the umbilical, under control of the IU pneumatic console. A low pressure switch monitors the sphere and, if the pressure falls below 1,000 psig, the ST-124-M3 stable platform is shut down to preclude damage to the gas bearing.

Output of the sphere is through a filter and a pressure regulator. The regulator reduces the sphere pressure to a level suitable for gas bearing lubrication. Pressure at the gas bearing is sampled and applied, as a control pressure to the regulator. This provides for a constant pressure across the gas bearing. From the main regulator, the gas flows through a heat exchanger where its temperature is stabilized, then through another filter and on to the gas bearing. Spent gas is then vented into the IU.

Figure 7-4 (Sheet 1 of 5)

INSTRUMENT UNIT EQUIPMENT LOCATIONS

Figure 7-4 (Sheet 2 of 5)

INSTRUMENT UNIT EQUIPMENT LOCATIONS

Figure 7-4 (Sheet 3 of 5)

INSTRUMENT UNIT EQUIPMENT LOCATIONS

Figure 7-4 (Sheet 4 of 5)

INSTRUMENT UNIT EQUIPMENT LOCATIONS

ST-124-M3
ELECTRONIC
ASSEMBLY

CONTROL
DISTRIBUTOR

GAS TEMP
SENSING
PROBE

REMOTE
DIGITAL
MULTIPLEXER
MODEL 410

18

IU COMMAND
DECODER

20

19

LAUNCH VEHICLE
DATA ADAPTER

PLATFORM
AC POWER
SUPPLY

LAUNCH VEHICLE
DIGITAL COMPUTER

ACCELEROMETER
SIGNAL
CONDITIONER

GAS BEARING
PRESSURE
REGULATOR

POSITION
IV

GAS BEARING
SOLENOID
VALVE

ACCELEROMETER
ASSEMBLY

GAS BEARING
HEAT
EXCHANGER

C-BAND
TRANSPONDER
NO 1

ST-124-M3
PLATFORM
ASSEMBLY

22

21

23

GN$_2$ 1000 PSI
SWITCH

GN$_2$ BEARING SUPPLY
(2 CU FT)

Figure 7-4 (Sheet 5 of 5)

THERMAL CONDITIONING SYSTEM FLOW DIAGRAM

Figure 7-5

HAZARDOUS GAS DETECTION

The hazardous gas detection system is used to monitor for the presence of hazardous gases in the IU and S-IVB stage forward compartment during vehicle fueling. The monitoring operation is continuous from the start of vehicle fueling to umbilical disconnect at liftoff.

The hazardous gas detection sampling equipment consists of four tubes which are open ended between panels 1 and 2, 7 and 8, 13 and 14, and 19 and 20. The tubes are connected to a quick disconect coupling on a single tube (see figure 7-8).

The hazardous gas detection equipment (GSE) extracts samples through the four tubes, and monitors the samples for the presence of hazardous gases.

ELECTRICAL POWER SYSTEMS

Primary flight power for the IU equipment is supplied by silver-zinc batteries at a nominal voltage level of 28 (± 2) vdc. During prelaunch operations, primary power is supplied by the GSE. Where ac power is required within the IU, it is developed by solid state dc to ac inverters. Power distribution within the IU is accomplished through power distributors which are, essentially, junction boxes and switching circuits.

BATTERIES

Silver-zinc primary flight batteries are installed in the IU during prelaunch operations. These batteries are physically and electrically identical, but each is connected to a separate bus in a power distributor. Flight components are connected to these busses in such a manner as to distribute the electrical load evenly between the batteries.

An attractive feature of the silver-zinc batteries is their high efficiency. Their ampere-hour rating is about four times as great as that of a lead-acid or nickle-cadmium battery of the same weight. The low temperature performance of the silver-zinc batteries is also substantially better than the others.

The battery characteristics are listed in figure 7-9.

POWER CONVERTERS

The IU electrical power systems contain a 56-volt power supply and a 5-volt measuring voltage supply.

56-Volt Power Supply

The 56-volt power supply furnishes the power required by the ST-124-M3 platform electronic assembly and the accelerometer signal conditioner. It is basically a dc to dc converter that uses a magnetic amplifier as a control unit. It converts the unregulated 28 vdc from the batteries to a regulated 56 vdc. The 56-volt power supply is connected to the platform electronic assembly through the power and control distributors.

Figure 7-6

PREFLIGHT AIR/GN₂ PURGE SYSTEM

Figure 7-7

HAZARDOUS GAS DETECTION SAMPLING EQUIPMENT LAYOUT

Figure 7-8

5-Volt Measuring Voltage Supply

The 5-volt measuring voltage supply converts unregulated 28 vdc to a closely regulated 5 (± .005) vdc for use throughout the IU measuring system. This regulated voltage is used primarily as excitation for measurement sensors (transducers), and as a reference voltage for inflight calibration of certain telemetry channels. Like the 56-volt supply, it is basically a dc to dc converter.

DISTRIBUTORS

The distribution system within the IU is comprised of the following:

1 Measuring distributor
1 Control distributor
1 Emergency Detection System (EDS) distributor
1 Power distributor
2 Auxiliary power distributors.

Measuring Distributor

The primary function of the measuring distributor is to collect all measurements that are transmitted by the IU telemetry system, and to direct them to their proper telemetry channels. These measurements are obtained from instrumentation transducers, functional components, and various signal and control lines. The measuring distributor also distributes the output of the 5-volt measuring voltage supply throughout the measuring system.

Through switching capabilities, the measuring distributor can change the selection of measurements monitored by the telemetry system. The switching function transfers certain measurements to channels which had been allotted to expended functions. If it were not for this switching, these channels would be wasted for the remainder of the flight.

IU BATTERY CHARACTERISTICS	
Type	Dry charge
Material	Alkaline silver-zinc
Cells	20 (with taps for selecting 18 or 19 cells if required to reduce high voltage)
Nominal voltage	1.5 per cell
Electrolyte	Potassium hydroxide (KOH) in demineralized water
Output voltage	+28 +2 vdc
Output current	35 amperes for a 10 hour load period (if used within 120 hours of activation)
Gross weight	165 pounds each

Figure 7-9

Control Distributor

The control distributor provides distribution of 28-volt power to small current loads, and distributes 56 vdc from the 56-volt power supply to the ST-124-M3 inertial platform assembly. The control distributor provides power and signal switching during prelaunch checkout for testing various guidance, control, and EDS functions, requested by the launch vehicle data adapter through the switch selector.

Emergency Detection System Distributor

The EDS distributor provides the only electrical link between the spacecraft and the LV. All EDS signals from the LV are routed to the logic circuits in the EDS distributor. EDS output signals from these logic circuits are then fed to the spacecraft and to the IU telemetry. Also, EDS signals from the spacecraft are routed back through the IU EDS logic circuits before being sent to the S-IVB, S-II, and S-IC stages.

Power Distributor

The power distributor provides primary distribution for all 28-volt power required by IU components. Inflight 28-volt battery power, or prelaunch ESE-supplied 28-volt power, is distributed by the power distributor.

The power distributor also provides paths for command and measurement signals between the ESE and IU components. The power distributor connects the IU component power return and signal return lines to the IU single point ground and to the umbilical supply return bus. These return lines are connected to the common bus in the power distributor, directly or indirectly, through one of the other distributors.

Silver-zinc batteries supply the inflight 28-vdc power. See figure 7-10 for a typical power distribution block diagram. Each battery is connected to a separate bus in the distributor. The flight components are connected to the busses in such a manner as to distribute the load evenly across the batteries.

Auxiliary Power Distributors

Two auxiliary power distributors supply 28 vdc power to small current loads. Both auxiliary power distributors receive 28 vdc from each of the battery busses in the power distributor, so that current loads on each of the batteries may be evenly distributed. Relays in the auxiliary power distributors provide power ON/OFF control for IU components both during the prelaunch checkout and while in flight. These relays are controlled by the ESE and by the switch selector.

IU GROUNDING

All IU grounding is referenced to the outer skin of the LV. The power system is grounded by means of hardwires routed from the power distributor COM bus to a grounding stud attached to the LV skin. All COM busses in the various other distributors are wired back to the COM bus in the power distributor. This provides for a single point ground.

Equipment boxes are grounded by direct metal-to-metal contact with cold plates or other mounting surfaces which are common to the LV skin. Most cabling shields are grounded to a COM bus in one of the distributors. However, where shielded cables run between equipment boxes, and not through a distributor, only one end of the shield would be grounded.

During prelaunch operations, the IU and GSE COM busses are referenced to earth ground. To ensure the earth ground reference until after all umbilicals are ejected, two single-wire grounding cables are connected to the IU below the umbilical plates. These are the final conductors to be disconnected from the IU.

Figure 7-10

EMERGENCY DETECTION SYSTEM

The EDS is one element of several crew safety systems. EDS design is a coordinated effort of crew safety personnel from several NASA centers.

The EDS senses initial development of conditions which could cause vehicle failure. The EDS reacts to these emergency situations in either of two ways. If breakup of the vehicle is imminent, an automatic abort sequence is initiated. If, however, the emergency condition is developing slowly enough, or is of such a nature that the flight crew can evaluate it and take action, only visual indications are provided to the flight crew. Once an abort sequence has been initiated, either automatically or manually, it is irrevocable and runs to completion.

The EDS is comprised of sensing elements, such as signal processing and switching circuitry, relay and diode logic circuitry, electronic timers and display equipment, all located in various places on the flight vehicle. Only that part of the EDS equipment located in the IU will be discussed here.

There are nine EDS rate gyros installed in the IU. Three gyros monitor each of the three axes (pitch, roll, and yaw) thus providing triple redundancy.

The control signal processor provides power to the nine EDS rate gyros, as well as receiving inputs from them. These inputs are processed and sent to the EDS distributor and to the flight control computer.

The EDS distributor serves as a junction box and switching device to furnish the spacecraft display panels with emergency signals if emergency conditions exist. It also contains relay and diode logic for the automatic abort sequence.

There is an electronic timer which is activated at liftoff, and which produces an output 30 seconds later. This output energizes relays in the EDS distributor which allows multiple engine shutdown. This function is inhibited during the first 30 seconds of launch.

Inhibiting of automatic abort circuitry is also provided by the LV flight sequencing circuits through the IU switch selector. This inhibiting is required prior to normal S-IC engine cutoff and other normal LV sequencing. While the automatic abort capability is inhibited, the flight crew must initiate a manual abort, if an angular-overrate or two-engine-out condition arises.

See Section III for a more complete discussion of emergency detection and procedures. Section III includes launch vehicle monitoring and control, EDS controls and abort modes and limits.

NAVIGATION, GUIDANCE AND CONTROL

The Saturn V launch vehicle is guided from its launch pad into earth orbit by navigation, guidance and control equipment located in the IU. An all inertial system, using a space stabilized platform for acceleration and attitude measurements, is utilized. A launch vehicle digital computer (LVDC) is used to solve guidance equations and a flight control computer (analog) is used for the flight control functions.

In the following discussions, the terms "navigation," "guidance," and "control" are used according to these definitions:

Navigation is the determination of the flight vehicle's present position and velocity from measurements made on board the vehicle.

Guidance is the computation of maneuvers necessary to achieve the desired flight path.

Control is the execution of the guidance maneuver by controlling the proper hardware.

Consider the block diagram of the overall Saturn V guidance and control subsystem shown in figure 7-11. The three-gimbal stabilized platform (ST-124-M3) provides a space-fixed coordinate reference frame for attitude control and for navigation (acceleration) measurements. Three integrating accelerometers, mounted on the gyro-stabilized inner gimbal of the platform, measure the three components of velocity resulting from vehicle propulsion. The accelerometer measurements are sent through the launch vehicle data adapter (LVDA) to the LVDC. In the LVDC, the accelerometer measurements are combined with the computed gravitational acceleration to obtain velocity and position of the vehicle.

The LVDA is the input/output device for the LVDC. It performs the necessary processing of signals to make these signals acceptable to the LVDC.

According to the guidance scheme (programmed into the computer), the maneuvers required to achieve the desired end conditions are determined by the LVDC. The instantaneous position and velocity of the vehicle are used as inputs. The result is the required thrust direction (guidance command) and the time of engine cutoff.

Control of the launch vehicle can be divided into attitude control and discrete control functions. For attitude control, the instantaneous attitude of the vehicle is compared with the desired vehicle attitude (computed according to the guidance scheme). This comparison is performed in the LVDC. Attitude correction signals are derived from the difference between the existing attitude angles (platform gimbal angles) and the desired attitude angles. In the flight control computer, these attitude correction signals are combined with signals from control sensors to generate the control commands for the engine actuators. The required thrust direction is obtained by gimbaling the engines in the propelling stage to change the thrust direction of the vehicle. In the S-IC and S-II, the four outboard engines are gimbaled to control roll, pitch, and yaw. Since the S-IVB stage has only one engine, an auxiliary propulsion system (APS) is used for roll control during powered flight. The APS provides complete attitude control during coast flight of the S-IVB/IU/spacecraft.

Guidance information stored in the LVDC (e.g., position, velocity) can be updated through the IU command system by data transmission from ground stations. The IU command system provides the general capability of changing or inserting information into the LVDC.

NAVIGATION, GUIDANCE & CONTROL SYSTEM BLOCK DIAGRAM

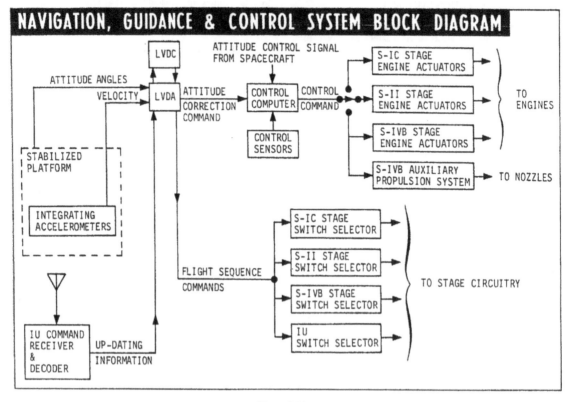

Figure 7-11

NAVIGATION SCHEME

Powered Flight

The basic navigation scheme is shown in figure 7-12. Gimbal resolvers supply platform position in analog form to the LVDA. An analog-to-digital converter in the LVDA converts the signal to the digital format required by the LVDC.

Platform integrating accelerometers sense acceleration components, and mechanically integrate them into velocity. The LVDA provides signal conditioning. Within the LVDC, initial velocity imparted by the spinning earth, gravitational velocity, and the platform velocities are algebraically summed together. This vehicle velocity is integrated by the LVDC to determine vehicle position.

Acceleration can be defined as the rate-of-change (derivative) of velocity. Velocity is the rate-of-change of position (distance). Velocity is the integral of acceleration, and position is the integral of velocity. Therefore, position is obtained by integrating acceleration twice.

Orbital Flight

During orbital coast flight, the navigational program continually computes the vehicle position, velocity, and acceleration from equations of motion which are based on vehicle conditions at the time of orbital insertion. In orbit,

navigation and guidance information in the LVDC can be updated by digital data transmission through the command and communications system.

Additional navigational computations are used in maintaining vehicle attitude during orbit. These computations establish a local vertical which is used as a reference for attitude control. The attitude of the vehicle roll axis will be maintained at 90 degrees with respect to the local vertical.

GUIDANCE COMPUTATIONS

The guidance function of the launch vehicle is accomplished by computing the necessary flight maneuvers to meet the desired end conditions of the flight (e.g., inserting the spacecraft into the desired trajectory). Guidance computations are performed within the LVDC by programmed guidance equations, which use navigation data and mission requirements as their inputs. These computations are actually a logical progression of computed formulas which have a guidance command as their solution. After the desired attitude has been determined by the "best path" program, the guidance computations might be reduced into the following equation: $\chi - \phi = \psi$ (See figure 7-13.)

where:

   χ is the desired attitude
   ϕ is the vehicle attitude
   ψ is the attitude error command

INSTRUMENT UNIT

Figure 7-12

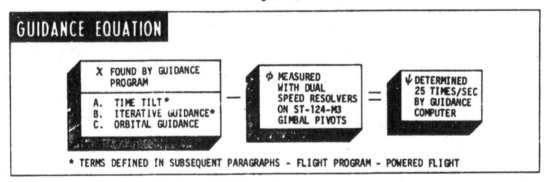

Figure 7-13

CONTROL SUBSYSTEM

The control subsystem (figure 7-14) is designed to control and maintain vehicle attitude by forming the steering commands used to control the engines of the active stage.

Vehicle attitude is achieved by gimbaling the four outboard engines of the S-IC stage, the four outboard engines of the S-II stage, or the single engine of the S-IVB stage. These engines are gimbaled by hydraulic actuators. Roll attitude control on the S-IVB stage cannot, of course, be controlled with a single engine. Therefore, roll control of the S-IVB stage is accomplished by the APS (see figure 7-15). During the coast period of the mission, the S-IVB APS will be used to control the vehicle attitude in all three axes.

The control system accepts guidance computations from the LVDC/LVDA guidance system. These guidance commands, which are actually attitude error signals, are then combined with measured data from the various control sensors. The resultant output is the command signal to the various engine actuators and APS nozzles.

The final computations (analog) are performed within the flight control computer. This computer is also the central switching point for command signals. From this point, the signals are routed to their associated active stages, and to the appropriate attitude control devices.

CONTROL SYSTEM COMPONENTS

Control Signal Processor

The control signal processor demodulates the ac signals from the control-EDS rate gyros into dc analog signals, required by the flight control computer. The control signal processor compares the output signals from the triple redundant gyros, and selects one each of the pitch, yaw, and roll signals for the flight control computer. The control signal processor supplies the control-EDS rate gyro package with the necessary control and reference voltages. EDS and DDAS rate gyro monitoring signals also originate within the control signal processor, thus accounting for the EDS portion of the control-EDS rate gyro name.

CONTROL SYSTEM BLOCK DIAGRAM

Figure 7-14

Flight Control Computer

The flight control computer is an analog computer which converts attitude correction commands (ψ) and angular change rates (ϕ) into APS thruster nozzle and/or engine actuator positioning commands.

Input signals to the flight control computer include:

1. Attitude correction commands (ψ) from the LVDC/LVDA or spacecraft

2. Angular change rates (ϕ) from the control-EDS rate gyro package, via the control signal processor.

Since one of the inputs to the flight control computer is from the control-EDS rate gyros, an excessive attitude correction command from the LVDC is limited within the flight control computer to protect the vehicle structure.

Output signals from the flight control computer include:

1. Command signals to the engine actuators (βc)

2. Command signals to the APS thruster nozzles (βc)

3. Telemetry outputs which monitor internal operations and functions.

ST-124-M3 Inertial Platform Assembly

The gimbal configuration of the ST-124-M3 offers unlimited freedom about the X & Y axes, but is limited to ±45 degrees about its Z axis (vehicle yaw at launch). See figure 7-16.

The gimbal system allows the inertial gimbal rotational freedom. Three single-degree-of-freedom gyroscopes have their input axes aligned along an orthogonal inertial coordinates system: X_1, Y_1, and Z_1 of the inertial gimbal. The signal generator, which is fixed to the output axis of each gyro, generates electrical signals proportional to torque disturbances. These signals are transmitted through the servo electronics which terminate in the gimbal pivot servotorque motors. The servo loops maintain the inner gimbal rotationally fixed in inertial space.

The inner gimbal has three, pendulous, integrating, gyroscopic accelerometers, oriented along the inertial coordinates X_1, Y_1, and Z_1. Each accelerometer measuring head contains a pendulous, single-degree-of-freedom gyro. The rotation of the measuring head is a measure of acceleration along the input axis of the accelerometer. Since acceleration causes the accelerometer shaft to be displaced as a function of time, the shaft position (with respect to a zero reference) is proportional to velocity, and the accelerometer is referred to as an integrating accelerometer.

Vehicle attitude is measured with respect to the inertial platform, using dual speed (32:1) resolvers located at the gimbal pivot points. The outputs of these angle encoders are converted into a digital count in the LVDA.

During prelaunch, the ST-124-M3 platform is held aligned to the local vertical by a set of gas bearing leveling pendulums. The pendulum output is amplified in the platform, and then transmitted to the ground equipment alignment amplifier. The alignment amplifier provides a signal to the torque drive amplifier, and then to the platform gyro torque generator. The vertical alignment system levels the platform to an accuracy of ±2.5 arc seconds.

The azimuth alignment is accomplished by means of a theodolite on the ground and two prisms on the platform, one fixed and one servo driven. The theodolite maintains the azimuth orientation of the movable prism, and the computer computes a mission azimuth, and programs the inner gimbal to that azimuth. The laying system has an accuracy of ±5 arc seconds.

At approximately liftoff minus seventeen seconds, the platform is released to maintain an inertial reference initiated at the launch point. At this time, the LVDC begins navigation, using velocity accumulations derived from the ST-124-M3 inertial platform.

Platform Electronic Assembly

The ST-124-M3 platform electronic assembly (PEA) contains the electronics, other than those located in the platform, required for the inertial gimbal and the accelerometer stabilization. Switching electronics for controlling platform system power and checkout functions are also located in the ST-124-M3 platform electronic assembly.

SATURN V ENGINES, ACTUATORS AND NOZZLE ARRANGEMENT

NOTES:

1. ALL SIGNAL ARROWS INDICATE POSITIVE VEHICLE MOVEMENTS.
2. VEHICLE PITCHES AROUND THE "Y" AXIS.
3. ENGINE ACTUATOR LAYOUTS SHOWN AS VIEWED FROM AFT END OF VEHICLE.
4. DIRECTIONS AND POLARITIES SHOWN ARE TYPICAL FOR ALL STAGES.
5. +β INDICATES ENGINE DEFLECTION REQUIRED TO CORRECT FOR POSITIVE VEHICLE MOVEMENT
6. CG = CENTER OF GRAVITY
   F = NOZZLES ON
   EXT = ACTUATOR EXTENDED
   RET = ACTUATOR RETRACTED
   β = THRUST VECTOR ANGULAR DEFLECTION

SC/LV COORDINATE RELATIONSHIPS

	SC	LV
POS I	-Z	+Z
POS II	+Y	-Y
POS III	+Z	-Z
POS IV	-Y	+Y

S-IC & S-II POLARITY TABLE

ACTUATOR NO.	ACTUATOR MOVEMENT		
	$+\phi_R$	$+\phi_Y$	$+\phi_P$
1-Y	RET	RET	
1-P	EXT		RET
2-Y	EXT	RET	
2-P	RET		EXT
3-Y	RET	EXT	
3-P	EXT		EXT
4-Y	EXT	EXT	
4-P	RET		RET

S-IVB POLARITY TABLE

ACTUATOR NO.	SIGNAL & ACTION			
	$+\psi_R$	$-\psi_R$	$+\psi_Y$	$+\psi_P$
1-Y			EXT	
1-P				RET
ENGINE NO.				
I_{IV}		F		
I_P				
I_{II}	F			
III_{II}		F		
III_P				
III_{IV}	F			

CONDITIONS DURING COAST

	$+\psi_R$	$-\psi_R$	$+\psi_Y$	$-\psi_Y$	$+\psi_P$	$-\psi_P$
I_{IV}		F		F		
I_P					F	
I_{II}	F		F			
III_{II}		F	F			
III_P						F
III_{IV}	F			F		

Figure 7-15

PLATFORM GIMBAL CONFIGURATION

Figure 7-16

The PEA includes the following circuitry:

1. Amplifiers, modulators, and stabilization networks for the platform gimbal and accelerometer servo loops

2. Relay logic for signal and power control

3. Amplifiers for the gyro and accelerometer pick-off coil excitation

4. Automatic checkout selection and test circuitry for servo loops

5. Control circuitry for the heaters and gas supply.

ST-124-M3 AC Power Supply

The ST-124-M3 platform ac power supply furnishes the power required to run the gyro rotors, and provides excitation for the platform gimbal synchros. It is also the frequency source for the resolver chain references and for gyro and accelerometer servo systems carrier.

The supply produces a three-phase, sine wave output which is fixed at 26 volts (rms) line-to-line at a frequency of 400 (± 0.01) Hertz. Three single-phase, 20-volt reference outputs (square wave) of 4.8 kHz, 1.92 kHz, and 1.6 kHz are also provided. With a normal input voltage of 28 vdc, the supply is capable of producing a continuous 250 va output.

Accelerometer Signal Conditioner

The accelerometer signal conditioner accepts the velocity signals from the accelerometer optical encoders and shapes them before they are passed on to the LVDA/LVDC.

Each accelerometer requires four shapers: a sine shaper and cosine shaper for the active channel, and a sine shaper and cosine shaper for the redundant channels. Also included are four buffer amplifiers for each accelerometer, one for each sine and cosine output.

Accelerometer outputs are provided for telemetry and ground checkout in addition to the outputs to the LVDA.

LV Digital Computer and LV Data Adapter

The LVDC and LVDA comprise a modern electronic digital computer system. The LVDC is a relatively high-speed computer with the LVDA serving as its input/output device. Any signal to or from the computer is routed through the LVDA. The LVDA serves as central equipment for interconnection and signal flow between the various systems in the IU. See figure 7-17 and 7-18 for LVDC and LVDA characteristics.

The LVDA and LVDC are involved in four main operations:

1. Prelaunch checkout

2. Navigation and guidance computations

3. Vehicle sequencing

4. Orbital checkout.

The LVDC is a general purpose computer which processes data under control of a stored program. Data is processed serially in two arithmetic sections which can, if so programmed, operate concurrently. Addition, subtraction, and logical extractions are performed in one arithmetic section, while multiplication and division are performed in the other.

The principal storage device is a random access, ferrite-core memory with separate controls for data and instruction addressing. The memory can be operated in either a simplex or duplex mode. In duplex operation, memory modules are operated in pairs with the same data being stored in each module. Readout errors in one module are corrected by using data from its mate to restore the defective location. In simplex operation, each module contains different data, which doubles the capacity of the memory. However, simplex operation decreases the reliability of the LVDC, because the ability to correct readout errors is sacrificed. The memory operation mode is program controlled. Temporary storage is provided by static registers, composed of latches, and by shift registers, composed of delay lines and latches.

Computer reliability is increased within the logic sections by the use of triple modular redundancy. Within this redundancy scheme, three separate logic paths are voted upon to correct any errors which develop.

LAUNCH VEHICLE DIGITAL COMPUTER CHARACTERISTICS	
ITEM	DESCRIPTION
Type	General purpose, digital, stored program
Memory	Random access, ferrite (torodial) core, with a capacity of 32,768 words of 28 bits each.
Speed	Serial processing at 512,000 bits per seconds
Word make-up	Memory = 28 bits Data = 26 bits plus 2 parity bits Instruction = 13 bits plus 1 parity bit
Programming	18 instruction codes 　10 arithmetic 　6 program control 　1 input/output 　1 store
Timing	Computer cycle = 82.03 μ sec. Bit time = 1.95 μ sec. Clock time = 0.49 μ sec.
Input/Output	External, program controlled

Figure 7-17

LAUNCH VEHICLE DATA ADAPTER CHARACTERISTICS	
ITEM	DESCRIPTION
Input/Output rate	Serial processing at 512,000 bits per second
Switch selector	8 bit input 15 bit output
Telemetry 　Command 　receiver	14 bits for input data
Data 　transmitter	38 data and identification bits plus validity bit and parity bit
Computer interface unit	15 bits address plus 1 data request bit 10 bits for input data plus 1 bit for data ready interrupt
Delay lines	3 four-channel delay lines for normal operation 1 four-channel delay line for telemetry operations
Output to launch computer	41 data and identification bits plus discrete outputs
Input from RCA-110 GCC	14 bits for data plus interrupt

Figure 7-18

FLIGHT PROGRAM

A flight program is defined as a set of instructions which controls the LVDC operation from seconds before liftoff until the end of the launch vehicle mission. These instructions are stored within the LVDC memory.

The flight program performs many functions during the launch vehicle mission. These functions include navigation, guidance, attitude control, event sequencing, data management, ground command processing, and hardware evaluation.

For purposes of discussion, the flight program is divided into five subelements: the powered flight major loop, the orbital flight program, the minor loop, interrupts, and telemetry.

The powered flight major loop contains guidance and navigation calculations, timekeeping, and all repetitive functions which do not occur on an interrupt basis. The orbital flight program consists of an executive routine concerned with IU equipment evaluation during orbit, and a telemetry time-sharing routine, to be employed while the vehicle is over receiving stations. In addition, in the orbital flight program, all navigation, guidance, and timekeeping computations are carried out on an interrupt basis, keyed to the minor loop. The minor loop contains the platform gimbal angle and accelerometer sampling routines and control system computations. Since the minor loop is involved with vehicle control, minor loop computations are executed at the rate of 25 times per second during the powered phase of flight. However, in earth orbit, a rate of only ten executions per second is required for satisfactory vehicle control.

PRELAUNCH AND INITIALIZATION

Until just minutes before launch, the LVDC is under control of the ground control computer (GCC). At approximately T-20 minutes, the GCC issues a prepare-to-launch (PTL) command to the LVDC. The PTL routine performs the following functions:

1. Executes an LVDC/LVDA self-test program and telemeters the results

2. Monitors accelerometer inputs, calculates the platform-off-level indicators, and telemeters accelerometer outputs and time

3. Performs reasonableness checks on particular discrete inputs and alerts

4. Interrogates the LVDC error monitor register

5. Keeps all flight control system ladder outputs zeroed, which keeps the engines in a neutral position for launch

6. Processes the GRR interrupt and transfers LVDC control to the flight program

7. Samples platform gimbal angles.

At T-22 seconds, the launch sequencer issues a GRR alert signal to the LVDC and GCC. At T-17 seconds a GRR interrupt signal is sent to the LVDC and GCC. With the receipt of this signal, the PTL routine transfers control of the LVDC to the flight program.

When the GRR interrupt is received by the LVDC, the following events take place:

1. The LVDC sets time base zero (T_0)

2. Gimbal angles and accelerometer values are sampled and stored for use by flight program routines

3. Time and accelerometer readings are telemetered

4. All flight variables are initialized

5. The GCC is signaled that the LVDC is under control of the flight program.

During the time period between GRR and liftoff, the LVDC begins to perform navigational calculations, and processes the minor loops. At T-8.9 seconds, engine ignition command is issued. At T-0, liftoff occurs, and a new time base (T_1) is initiated.

POWERED FLIGHT MAJOR LOOP

The major loop contains the navigation and guidance calculations, timekeeping, and other repetitive operations of the flight program. Its various routines are subdivided by function. Depending upon mode of operation and time of flight, the program will follow the appropriate sequence of routines.

The accelerometer processing routine accomplishes two main objectives: it accumulates velocities as measured by the platform, and detects velocity measurement errors through "reasonableness" tests.

The boost navigation routine combines gravitational acceleration with measured platform data to compute position and velocity.

The "pre-iterative" guidance mode, or "time-tilt" guidance program, is that part of the flight program which performs from GRR until the end of the SIC burn. The guidance commands issued during the time-tilt phase are functions of time only. This phase of the program is referred to as open loop guidance, since vehicle dynamics do not affect or influence the guidance commands. When the launch vehicle has cleared the mobile launcher, the time-tilt program first initiates a roll maneuver to align the vehicle with the proper azimuth. After this command, roll and yaw commands remain at zero, and the vehicle is gradually pitched about the vehicle's Y axis to its predetermined boost heading. Rate limiting of the output commands prevents the flight control system from maneuvering the LV at rates which exceed safe limits.

The iterative guidance mode (IGM) routine, or "path adaptive" guidance, commences after second stage ignition, and continues until the end of S-IVB first burn. Cutoff occurs when the velocity required for earth orbit has been reached. IGM is used again during S-IVB second burn. IGM is based on optimizing techniques, using the calculus of variations to determine a minimum propellant flight path which satisfies mission requirements. Since the IGM considers vehicle dynamics, it is referred to as closed loop guidance.

INTERRUPTS

An interrupt routine permits interruption of the normal program operation to free the LVDC for priority work, and may occur at any time within the program. When an interrupt occurs, the interrupt transfers LVDC control to a special subroutine which identifies the interrupt source, performs the necessary subroutines, and then returns to the point in the program where the interrupt occurred. Figure 7-19 is a list of interrupts in the order of decreasing priority.

TELEMETRY ROUTINE

A programmed telemetry feature is also provided as a method of monitoring LVDC and LVDA operations. The telemetry routine transmits specified information and data to the ground via IU telemetry equipment. In orbit, telemetry data must be stored at times when the vehicle is not within range of a ground receiving station. This operation is referred to as data compression. The stored data is transmitted on a time-shared basis with real-time telemetry when range conditions are favorable.

DISCRETE BACKUPS

Certain discrete events are particularly important to the flight program, since they periodically reset the computer time base which is the reference for all sequential events. These significant time base events are:

T_1 Liftoff (LO)
T_2 S-IC center engine cutoff (CECO)
T_3 S-IC outboard engine cutoff (OECO)
T_4 S-II cutoff
T_5 S-IVB cutoff (end boost phase)
T_6 S-IVB restart preparations & second burn
T_7 S-IVB cutoff
T_8 S-IVB propellant dump sequence

Since switch selector outputs are a function of time (relative to one of the time bases), no switch selector output could be generated if one of the discrete signals were missed. A backup routine is provided to circumvent such a failure. The discrete backup routine simulates these critical signals if they do not occur when expected.

In the cases of the backup routines for LO and CECO, special routines are established as a double safety check. In both cases, motion as well as time are confirmed before a backup discrete is used. For LO, the backup routine is entered 17.5 seconds after GRR. If the vertical acceleration exceeds 6.544 ft/sec^2 for four computation cycles, the vehicle is assumed to be airborne and the liftoff discrete is issued. For CECO, an assurance is made that an on-the-pad firing of the S-II stage cannot occur if T_1 is accidentally set. Before T_2 can be initiated, velocity along the downrange axis is tested for a minimum of 500 m/sec.

Refer to MODE AND SEQUENCE CONTROL, Section VII, for a discussion of the discrete time bases T_1 through T_8.

The execution time for any given major loop, complete with minor loop computations and interrupts, is not fixed. The average execution time for any given major loop in powered flight, complete with minor loop computation and interrupt processing, is called the normal computation cycle for that mode. The computation cycle is not fixed for two reasons.

First, the various flight modes of the program have different computation cycle lengths. Second, even in a given flight mode, the uncertainties of discrete and interrupt processing, and the variety of possible paths in the loop preclude a fixed computation cycle length.

MODE AND SEQUENCE CONTROL

Mode and sequence control involves most of the electrical/electronic systems in the launch vehicle. However, in this section the discussion will deal mainly with the switch selector and associated circuitry.

The LVDC memory contains a predetermined number of sets of instructions which, when initiated, induce portions of the launch vehicle electrical/electronic systems to operate in a particular mode. Each mode consists of a predetermined sequence of events. The LVDC also generates appropriate discrete signals such as engine ignition, engine cutoff, and stage separation.

Mode selection and initiation can be accomplished by an automatic LVDC internal command, an external command from ground checkout equipment or IU command system, or by the flight crew in the spacecraft.

The flexibility of the mode and sequence control scheme is such that no hardware modification is required for mode and flight sequence changes. The changes are accomplished by changing the instructions and programs in the LVDC memory.

SWITCH SELECTOR

Many of the sequential operations in the launch vehicle that are controlled by the LVDC are performed through a switch selector located in each stage. The switch selector decodes digital flight sequence commands from the LVDA/LVDC, and activates the proper stage circuits to execute the commands. The outputs of the switch selector drive relays, either in the units affected or in the stage sequencer.

INTERRUPTS	
Decreasing Priority	Function
1	Minor loop interrupt
2	Switch selector interrupt
3	Computer interface unit interrupt
4	Temporary loss of control
5	Command receiver interrupt
6	Guidance reference release
7	S-II propellant depletion/engine cutoff
8	S-IC propellant depletion/engine cutoff "A"
9	S-IVB engine out "B"
10	Program re-cycle (RCA-110A interrupt)
11	S-IC inboard engine out "A"
12	Command LVDA/RCA-110A interrupt

Figure 7-19

Each switch selector can activate, one at a time, 112 different circuits in its stage. The selection of a particular stage switch selector is accomplished through the command code. Coding of flight sequence commands and decoding by the stage switch selectors reduces the number of interface lines between stages, and increases the flexibility of the system with respect to timing and sequence. In the launch vehicle, which contains four switch selectors, 448 different functions can be controlled, using only 28 lines from the LVDA. Flight sequence commands may be issued at time intervals of 100 milliseconds.

To maintain power isolation between vehicle stages, the switch selectors are divided into sections. The input sections (relay circuits) of each switch selector receive their power from the IU. The output sections (decoding circuitry and drivers) receive their power from the stage in which the switch selector is located. The inputs and outputs are coupled together through a diode matrix. This matrix decodes the 8-bit input code, and activates a transistorized output driver, thus producing a switch selector output.

The output signals of the LVDA switch selector register, with the exception of the 8-bit command, are sampled at the control distributor in the IU and sent to IU PCM telemetry. Each switch selector also provides three outputs to the telemetry system within its stage.

The switch selector is designed to execute flight sequence commands given by the 8-bit code or by its complement. This feature increases reliability and permits operation of the system, despite certain failures in the LVDA switch selector register, line drivers, interface cabling, or switch selector relays.

The flight sequence commands are stored in the LVDC memory, and are issued according to the flight program. When a programmed input/output instruction is given, the LVDC loads the 15-bit switch selector register with the computer data.

The switch selector register, bits 1 through 8, represents the flight sequence command. Bits 9 through 13 select the switch selector to be activated. Bit 14 resets all the relays in the switch selectors in the event data transfer is incorrect, as indicated by faulty verification information received by the LVDA. Bit 15 activates the addressed switch selector for execution of the command. The switch selector register is loaded in two passes by the LVDC: bits 1 through 13 on the first pass, and either bit 14 or bit 15 on the second pass, depending on the feedback code. The LVDA/LVDC receives the complement of the code after the flight sequence command (bits 1 through 8) has been picked up by the input relays of the switch selector. The feedback (verification information) is returned to the LVDA, and compared with the original code in the LVDC. If the feedback agrees, the LVDC/LVDA sends a read command to the switch selector. If the verification is not correct, a reset command is given (forced reset), and the LVDC/LVDA reissues the 8-bit command in complement form.

Figure 7-20 illustrates the Saturn V switch selector functional configuration. All switch selector control lines are connected through the control distributor in the IU to the LVDC and the electrical support equipment.

The LVDC switch selector interconnection diagram is shown in figure 7-21. All connections between the LVDA and the switch selectors, with the exception of the stage select inputs, are connected in parallel.

OPERATION SEQUENCE

The Saturn V operation sequence starts during the prelaunch phase at approximately T-24 hours, when the electrical power from the ground support equipment is applied to all stages of the launch vehicle. During this time, the sequencing is controlled from the launch control center/mobile launcher complex, utilizing both manual and automatic control to check out the functions of the entire launch vehicle. After the umbilicals are disconnected, the sequencing is primarily controlled by the flight program within the LVDC.

Since flight sequencing is time phased, the sequencing operation is divided into eight primary time bases. Each time base is related to a particular flight event. Four alternate time bases have also been provided for. These time bases are defined in the following paragraphs.

Time Base No. 1 (T_1)

T_1 is initiated by a liftoff signal, provided by deactuation of the liftoff relay in the IU at umbilical disconnect. However, as a safety measure, the LVDC will not recognize the liftoff signal and start T_1 prior to receiving guidance reference release (GRR) plus 16.0 seconds.

A backup method for starting T_1 is provided should the LVDC fail to receive or recognize the liftoff signal. If T_1 is not initiated within 17.5 seconds after GRR, the LVDC will monitor the vertical accelerometer. If a significant positive acceleration (in excess of 1 g) exists, the LVDC assumes liftoff has occurred and begins T_1. A compensating time adjustment is made by the LVDC.

No "negative backup" (i.e., provisions for the LVDC to return to prelaunch conditions) is provided because, in the event T_1 began by error, the launch vehicle could safely complete T_1 on the pad without catastrophic results.

Time Base No. (T_2)

The S-IC center engine is cut off by the LVDC, through the S-IC switch selector, at T_1 + 2 minutes 14.6 seconds. At T_1 + 2 minutes 14.7 seconds, the LVDC monitors the downrange accelerometer. If sufficient downrange velocity exists, the LVDC will start T_2.

Use of the downrange velocity reading provides a safeguard against starting T_2 on the pad, should T_1 be started without liftoff. Furthermore, if T_2 is not established, no subsequent time bases can be started. This ensures a safe vehicle, requiring at least one additional failure to render the vehicle unsafe on the pad.

Time Base No. 3 (T_3)

T_3 is initiated at S-IC outboard engine cutoff by either of two redundant outboard engines cutoff signals. However, the LVDC must arm outboard engines propellant depletion cutoff prior to starting T_3. Outboard engines propellant depletion cutoff is armed several seconds prior to calculated outboard engines cutoff.

Time Base No. 4 (T_4)

After arming S-II lox depletion cutoff sensors, the LVDC

SWITCH SELECTOR FUNCTIONAL CONFIGURATION

NOTES: SIGNAL RETURN LINES FROM THE SWITCH SELECTORS, THROUGH THE CONTROL
DISTRIBUTOR, TO THE LVDA ARE NOT SHOWN IN THIS FIGURE.
THE LETTERS USED TO LABEL INTERSTAGE CONNECTIONS BETWEEN UNITS ARE
NOT ACTUAL PIN OR CABLE CONNECTORS. THE LETTER CODE IS DENOTED
BELOW:

a = 8-DIGIT COMMAND (8 LINES)
b = FORCE RESET (REGISTER) (1 LINE + 1 REDUNDANT LINE)
c = REGISTER VERIFICATION (8 LINES)
d = READ COMMAND (1 LINE + 1 REDUNDANT LINE)
e = ⎫
f = ⎬   STAGE SELECT LINES
g = ⎬   (1 LINE + 1 REDUNDANT LINE)
h = ⎭
j = b, c, d, e, f, g, AND h TO IU TELEMETRY
k = REGISTER TEST ⎫
    ZERO INDICATE ⎬ TO STAGE TELEMETRY
    SW SEL OUTPUT ⎭   (1 LINE EACH)
m = +28 VDC FROM THE INSTRUMENT UNIT

Figure 7-20

Figure 7-21

initiates T$_4$ upon receiving either of two signals: S-II engines cutoff, or S-II engines out. The S-II engines cutoff signal is the primary signal for starting T$_4$. The S-II engines out signal from the thrust OK circuitry is a backup.

As a safeguard against trying to separate the S-II stage with the thrust of the engines present, a redundant S-II engines cutoff command is issued by the LVDC at the start of T$_4$.

Alternate Time Base No. 4a (T$_{4a}$)

T$_{4a}$ is programmed for use in early staging of the S-IVB stage. This time base is initiated by the LVDC upon receiving either of two signals: spacecraft initiation of S-II/S-IVB separation "A", or spacecraft initiation of S-II/S-IVB separation "B". Starting of T$_{4a}$ is inhibited until T$_3$ + 1.4 seconds.

Time Base No. 5 (T$_5$)

T$_5$ is initiated by the deactuation of the S-IVB thrust OK pressure switches at S-IVB cutoff. The LVDC starts T$_5$ after receiving any two of four functions monitored by the LVDC. The functions are: S-IVB engine out "A"; S-IVB engine out "B"; S-IVB velocity cutoff, which is issued by the LVDC; and/or loss of thrust, determined by LVDC using accelerometer readings.

A redundant S-IVB cutoff command is issued at the start of T$_5$ as a safeguard against having started time base 5 with the thrust of the S-IVB engine present.

Time Base No. 6 (T$_6$)

After a predetermined time in T$_5$, T$_6$ is initiated by the LVDC upon solving the restart equation.

The LVDC also has the capability to initiate T$_6$ by use of a target update. When this is used, solving of the restart equation is inhibited.

However, the above starting logic for T$_6$ is inhibited by the following:

1. Translunar Injection signal from the SC

2. Guidance Failure "A" or "B" signals

3. DC Command for TD&E enable

Also, if the DCS commanded for TD&E enable is received by the LVDC, the starting of T$_6$ is permanently inhibited.

In the event of Guidance Failure and SC Control of Saturn, T$_6$ is initiated by the LVDC after T$_5$ + 100.0 seconds by the S-IVB Ignition Sequence Start signal from the SC. This logic is inhibited only by the DCS command for TD&E enable.

Alternate Time Base No 6$_a$ (T$_{6a}$)

T$_{6a}$ is programmed for use should the oxygen-hydrogen burner malfunction between the times T$_6$ + 48 seconds and T$_6$ +5 minutes 41.3 seconds. This alternate time base is initiated by the LVDC upon receiving an oxygen-hydrogen burner malfunction signal from the S-IVB stage. Upon completion of T$_{6a}$, the LVDC returns to T$_6$.

Alternate Time Base No. 6b (T$_{6b}$)

T$_{6b}$ is programmed for use if the O$_2$/H$_2$ burner malfunctions between T$_6$ + 5 minutes 41.3 seconds and T$_6$ + 8 minutes 16.7 seconds. This alternate time base is initiated by the LVDC upon receiving an O$_2$/H$_2$ burner malfunction signal from the S-IVB.

Alternate Time Base No. 6c (T$_{6c}$)

T$_{6c}$ is programmed for use should a failure occur which would require a delay in the S-IVB restart attempt. The TRANSLUNAR INJECTION INHIBIT signal from the spacecraft will be required by the LVDC before this alternate time base will be initiated.

The LVDC is programmed to look for the Translunar Injection (TLI) Inhibit signal at T$_6$ + 41 seconds, at which time T$_{6c}$ will be initiated if the TLI inhibit signal is present.

The LVDC is also programmed to look for the TLI inhibit signal at T$_6$ + 8 minutes 17.3 seconds and then once per computer cycle (once per second) between T$_6$ + 8 minutes 17.3 seconds and T$_6$ + 9 minutes 20 seconds. If the TLI inhibit signal is present the LVDC will initiate T$_{6c}$.

Upon completion of T$_{6c}$ the LVDC will return to T$_5$ updated by the time elapsed in T$_6$ and T$_{6c}$.

Time Base No. 7 (T$_7$)

After a predetermined time, sufficient to allow the S-IVB engine to establish thrust OK, the LVDC starts T$_7$ after receiving any two of four functions monitored by the LVDC. The functions are: S-IVB engine out "A"; S-IVB engine out "B"; S-IVB engine cutoff, which is issued by the LVDC; and/or loss of thrust, determined by LVDC using accelerometer readings.

As a safeguard against starting T$_7$ with the thrust of the S-IVB engine present, a redundant S-IVB engine cutoff command is issued at the start of T$_7$.

Time Base No. 8 (T$_8$)

The start of T$_8$ is program inhibited within the LVDC. This program inhibit must be removed by DCS command. However, the LVDC will not accept the DCS command to remove the inhibit (S-IVB propellant dump inhibit) until 8 minutes after the DCS evasive maneuver command has been received. The LVDC will not accept the DCS evasive maneuver command until T$_7$ + 2 hours. The LVDC will start T$_8$ subsequent to T$_7$ + 2 hours if the program inhibit is removed.

MEASUREMENTS AND TELEMETRY

The instrumentation within the IU consists of a measuring subsystem, a telemetry subsystem, and an antenna subsystem. This instrumentation is for the purpose of monitoring certain conditions and events which take place within the IU, and for transmitting monitored signals to ground receiving stations. Telemetry data is used on the ground for the following purposes:

1. Before launch, for checkout of the launch vehicle

2. During vehicle flight, for immediate determination of vehicle condition, and for verification received by the IU command system

3. Postflight scientific analysis of the mission.

MEASUREMENTS

The requirement for measurements of a wide variety has dictated the use of many types of transducers at many different locations. However, a discussion of each transducer type is beyond the scope of this manual. The parameters measured include such things as acceleration, angular velocity, flow rate, position, pressure, temperature, voltage, current, frequency, etc.

Conditioning of measured signals is accomplished by amplifiers or converters located in measuring racks. There are four measuring racks in the IU, and 20 signal conditioning modules in each. Each signal conditioning module contains, in addition to its conditioning circuitry, two relays and circuitry to simulate its transducers at both their high range and low range extremities. These relays, and transducer simulation circuitry, are used for prelaunch calibration of the signal conditioners.

Conditioned signals are routed to their assigned telemetry channel by the measuring distributors. Switching functions connect different sets of measurements to the same telemetry channels during different flight periods. These switching functions, controlled from the ground through the umbilical, connect measurements not required during flight to digital data acquisition system channels for ground checkout, and return the channels to flight measurements after checkout.

TELEMETRY

The function of the telemetry system is to format and transmit measurement signals received from the measuring distributor.

The approximately 200 measurements made on the IU are transmitted via two telemetry links. The two modulation techniques used are:

1. Pulse Code Modulation/Frequency Modulation (PCM/FM)

2. Frequency Modulation/Frequency Modulation (FM/FM)

The PCM/FM data is transmitted over a VHF link and over the CCS transponder. The FM/FM data is transmitted over a separate VHF link.

Multiplexing

In order for the two IU telemetry links to handle approximately 200 measurements, these links must be "shared". By proper multiplexing, it is possible to transmit several different signals simultaneously from one telemetry system. Both frequency sharing and time sharing multiplexing techniques are used to accomplish this. Refer to figure 7-22 for a block diagram of the IU telemetry system.

Two Model 270 multiplexers (MUX-270) are used in the IU telemetry system. The MUX-270 is a time sharing multiplexer. Each one operates as a 30 x 120 (30 primary channels, each sampled 120 times per second) multiplexer with provisions for submultiplexing individual primary channels to form ten subchannels, each sampled at 12 times per second. Twenty-seven of the 30 primary channels are used for analog data, while the remaining three are used for references. Ten-channel submultiplexer modules, which plug into the MUX-270, can be used to submultiplex any primary data channel, providing a 10 to 1 increase in the quantity of channels, in exchange for a 10 to 1 decrease in sampling rates. Any proportion of the 23 data channels can be submultiplexed or sampled at the 120 per second rate.

The MUX-270 also has an integral calibration generator for inflight calibration capability. Upon command, the calibration generator seeks the next available master frame, and applies a sequence of five calibration voltages to all data channels. Each level is sustained for one master frame, and approximately 400 milliseconds are required for the full sequence.

The Model 410 remote digital multiplexer (RDM-410) accepts ten 10-bit parallel words, and transfers this data to the Model 301 PCM/DDAS according to a programmed format.

Low level conditioned analog signals are fed to subcarrier oscillators (SCO). The Model B1 SCO has a capacity of 28 continuous data channel inputs. Each input is applied to a separate channel within the SCO, and each channel produces a different output frequency. These output frequencies are combined within the SCO assembly, and the composite signal is used to frequency modulate an FM-RF assembly producing the FM/FM telemetry signals.

The PCM/FM system performs a dual function. During flight, it serves as a telemetry link, and during prelaunch checkout, it serves as an IU interface with the digital GSE. PCM techniques provide the high degree of accuracy required for telemetering certain signal types. The PCM-301 unit accepts analog inputs from MUX-270 or RDM-410 units, or direct inputs in digital form. All inputs are digitized and encoded. Output of the PCM-301 unit is a serial train of digital data which modulates the PCM-RF transmitter.

All of the RF assemblies are essentially the same. All use

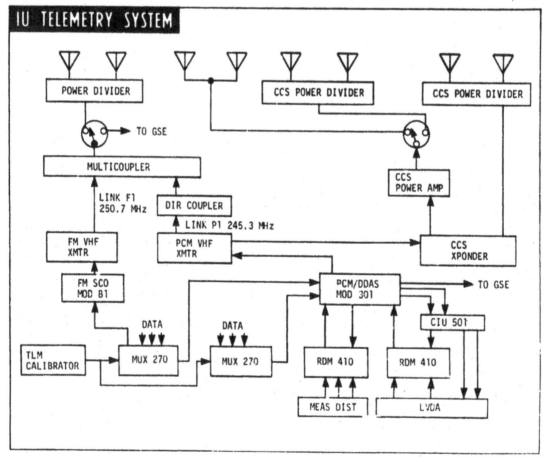

Figure 7-22

combinations of solid state and vacuum tube electronics. Frequency outputs of each unit are, of course, different, and are applied to the antenna subsystem.

RADIO COMMAND SYSTEM

COMMAND COMMUNICATIONS SYSTEM (CCS)

The CCS provides for digital data transmission from ground stations to the LVDC. This communications link is used to update guidance information, or to command certain other functions through the LVDC. Command data originates in the Mission Control Center, and is sent to remote stations of the MSFN for transmission to the launch vehicle.

At the time of spacecraft separation from the IU/S-IVB, the IU CCS transmitter will be commanded off for a short period of time, to preclude interference with the spacecraft S-band transponder. After adequate separation of the spacecraft and IU/S-IVB, the IU CCS transmitter will be commanded on again, to provide for pseudo-random noise (PRN) turnaround ranging, to facilitate IU/S-IVB tracking.

The CCS equipment located in the IU consists of:

1. Antenna systems
   a. transmitting and receiving
   b. directional and omni

2. Antenna switching elements
   a. coaxial switches
   b. hybrid rings
   c. power divider

3. S-band transponder:
   a phase coherent receiver-transmitter

4. Command decoder, which precludes unauthorized command data entry.

Figure 7-23 is a block diagram of the overall CCS. Command messages are transmitted from the unified S-band ground stations on a carrier frequency of 2101.8 MHz, modulated by a subcarrier of 70 kHz, which is modulated by a digital message. The transmitted message is received by the airborne transponder, where demodulation is accomplished. The resulting digital message is passed on to the command decoder, where it is checked for authenticity before being passed to the LVDC. The IU PCM telemetry system verifies receipt of the message.

The LVDC is programmed to receive two types of command messages from the command decoder: command words and data words.

The LVDC can be programmed to recognize as many as 26 different mode command words. Many of these command words are common to all flights while others are programmed only for particular missions. Common mode commands include:

1. time base update
2. navigational update
3. execute switch selector routine
4. telemeter memory contents
5. terminate command routine.

Data words, as the name implies, contain data to supplement mode commands. The number of data words varies with the mode command involved. For example, a time base update requires only one data word while a navigation update requires more than thirty data words.

SATURN TRACKING INSTRUMENTATION

The purpose of radio tracking is the determination of the vehicle's trajectory. Tracking data is used for mission control, range safety, and postflight evaluation of vehicle performance.

The Saturn V IU carries two C-band radar transponders. The tracking of the launch vehicle may be divided into four phases: powered flight into earth orbit, orbital flight, injection into mission trajectory, and coast flight after injection.

Continuous tracking is required during powered flight into earth orbit. Because of the long burning time (approximately 11 minutes) of the three-stage launch vehicle, the end of the powered flight phase cannot be covered sufficiently from land-based tracking stations. Therefore, tracking ships will be located in the Atlantic to obtain the tracking data, during insertion, which is required for orbit determination. The number of stations which can "see" the vehicle depends on the launch azimuth.

In addition, the launch vehicle will be tracked from S-band stations at Cape Kennedy and on the Atlantic tracking ships. These stations have dual tracking capability: i.e., they can simultaneously track the two S-band transponders on the vehicle, one in the IU and the other in the Apollo spacecraft. The S-band station on Bermuda has only a single capability, and will track the Apollo spacecraft transponder. Refer to Radio Command Systems for additional information on the S-band equipment.

During orbital flight, tracking is accomplished by S-band stations of the MSFN and by C-band radar stations. The S-band stations, including the Deep Space Instrumentation Facility, can track the Apollo spacecraft to the moon, and will also be involved in tracking after injection. Tracking information collected during orbital flight may be used to update the Saturn guidance before injection into mission trajectory.

C-BAND RADAR

The function of the C-band radar transponder is to increase the range and accuracy of the radar ground stations equipped with AN/FPS-16, and AN/FPQ-6 radar systems. C-band radar stations at the Kennedy Space Center, along the Atlantic Missile Range, and at many other locations around the world, provide global tracking capabilities. Two C-band radar transponders are carried in the IU to provide radar tracking capabilities independent of the vehicle attitude. This arrangement is more reliable than the antenna switching circuits necessary if only one transponder were used.

The transponder consists of a single, compact package. Major elements include an integrated RF head, an IF amplifier, a decoder, overinterrogation protection circuitry, a fast recovery solid-state modulator, a magnetron, a secondary power supply, and transducers for telemetry channels. The complete unit weighs 5.5 pounds, and has a volume of only 100 cubic inches.

The transponder receives coded or single-pulse interrogation from ground stations, and transmits a single-pulse reply in the same frequency band.

Six conditioned telemetry outputs are provided: input signal level, input PRF, temperature, incident power, reflected power, and reply PRF.

The characteristics of the C-band radar transponder are given in figure 7-24.

GROUND SUPPORT EQUIPMENT

The IU, because of its complex nature, requires the services of many types of GSE (mechanical, pneudraulic, electrical, electronic) and personnel. This section of the manual is limited to a very brief description of the IU GSE.

There are three primary interfaces between the IU and its GSE. One is the IU access door, used during prelaunch preparations for battery installation, ordnance servicing, servicing IU equipment, S-IVB forward dome and LEM servicing. The second interface is the umbilical, through which the IU is furnished with ground power, purging air/GN$_2$ methanol-water for environmental control, and hardwire links with electrical/electronic checkout equipment. The third interface is the optical window, through which the guidance system ST-124-M3 stable platform is aligned.

IU ACCESS DOOR

The structure of the IU consists of three 120-degree segments of aluminum honeycomb sandwich, joined to form a cylindrical ring. After assembly of the IU, a door assembly provides access to the electronic equipment inside the structure. This access door has been designed to act as a load supporting part of the structure in flight.

Work platforms, lights, and air-conditioning are used inside the IU to facilitate servicing operations. When the spacecraft is being fueled through the IU access door, a special protective cover is installed inside the IU to protect components from any possible volatile fuel spillage.

Approximately 20 hours before launch, the IU flight batteries, each weighing 165 pounds, are activated in the battery shop and installed in the IU through the access door.

At approximately T-6 hours, the service equipment is removed and the access door is secured.

IU UMBILICAL

The physical link between the IU and the GSE is through the umbilical connection, located adjacent to the access door. The umbilical is made up of numerous electrical connectors,

Figure 7-23

two pneudraulic couplings and an air conditioning duct. The electrical connectors provide ground power and the electrical/electronic signals necessary for prelaunch checkout of the IU equipment. The pneudraulic couplings provide for circulation of GSE supplied methanol-water coolant fluid for the IU/S-IVB ECS. The air conditioning duct provides for compartment cooling air or purging GN_2.

The umbilical is retracted at liftoff, and a spring loaded door on the IU closes to cover the connectors.

OPTICAL ALIGNMENT

The IU contains a window through which the ST-124-M3 stable platform has its alignment checked and corrected by a theodolite located in a hut on the ground and a computer feedback loop. By means of this loop, the launch azimuth can be monitored, updated and verified to a high degree of accuracy.

IU/SLA INTERFACE

MECHANICAL INTERFACE

The IU and spacecraft-LM adapter (SLA) are mechanically aligned with three guide pins and brackets as shown in figure 7-25. These pins facilitate the alignment of the close tolerance interface bolt holes, as the two units are joined during vehicle assembly. Six bolts are installed around the circumference of the interface and sequentially torqued, using a special MSFC designed wrench assembly. These six bolts secure the IU/SLA mechanical interface. (See figure 7-26.)

ELECTRICAL INTERFACE

The electrical interface between the IU and spacecraft consists of three 61 pin connectors. (See figure 7-27.) The definition and function of each connector is presented in the following paragraphs.

C-BAND TRANSPONDER CHARACTERISTICS	
Receiver characteristics	
Frequency (tunable externally)	5400 to 5900 MHz (set to 5690 $\pm$2 MHz)
Frequency stability	$\pm$2.0 MHz
Bandwidth (3 db)	10 MHz
Off-frequency rejection	50 db image; 80 db minimum, 0.15 to 10,000 MHz
Sensitivity (99% reply)	-65 dbm over entire frequency range and all environments
Maximum input signal	+20 dbm
Interrogation code	Single or double pulse
Pulse width	0.2 to 5.0 μsec (single pulse), 0.2 to 1.0 μsec (double pulse)
Pulse spacing	Continuously settable between 5 and 12 μsec (set to 8 $\pm$0.05 μsec)
Decoder limits	$\pm$0.25 μsec accept, $\pm$ 0.85 μsec reject (5 to 12 μsec)
Transmitter characteristics	
Frequency (tunable externally)	5400 to 5900 MHz (set to 5765 $\pm$2 MHz)
Peak power output	400 watts minimum, 700 watts nominal
Pulse width	1.0 $\pm$ 0.1 μsec
Pulse jitter	0.020 μsec maximum for signals above -55 dbm
Pulse rise time (10% to 90%)	0.1 μsec maximum
Duty cycle	0.002 maximum
VSWR of load	1.5:1 maximum
Pulse repetition rate	10 to 2000 pps; overinterrogation protection allows interrogation at much higher rates with count-down; replies during overinterrogation meet all requirements
Transponder characteristics	
Recover time	50 μsec single pulse, 62 μsec double pulse maximum for input signal levels differing by up to 65 db (recovers to full sensitivity with no change in transmitter reply power or frequency with multiple radars interrogating simultaneously)
Fixed delay	Settable 2 $\pm$ 0.1 and 3.0 to 0.01 μsec (set to 3.0 $\pm$ 0.01 μsec)
Delay variation with signal level	50 nanoseconds maximum from -65 dbm to 0 dbm
Power requirements	24 to 30 volts
Primary current drain	0.7 ampere standby; 0.9 ampere at 1000 pps
Weight	5.5 lbs

Figure 7-24

IU/Spacecraft Interface Connector J-1

This connector provides lines for power, control, indication circuitry and EDS circuitry.

IU/Spacecraft Interface Connector J-2

This connector provides lines for power, control and indications for the Q-ball circuitry and the EDS circuitry.

IU/Spacecraft Interface Connector J-3

This connector provides lines for power, control, and indication circuitry and EDS circuitry.

MAJOR DIFFERENCES

There are no major differences between Saturn V IU-6, IU-7, IU-8, and IU-9.

Figure 7-25

Figure 7-26

Figure 7-27

TABLE OF CONTENTS

LAUNCH COMPLEX 39

Launch Complex 39 (LC-39), Kennedy Space Center, Florida, provides all the facilities necessary to the assembly, checkout, and launch of the Apollo/Saturn space vehicle. The vehicle assembly building (VAB) provides a controlled environment in which the vehicle is assembled and checked out on a mobile launcher (ML). The space vehicle and the launch structure are then moved as a unit by the crawler-transporter to the launch site, where vehicle launch is accomplished after propellant loading and final checkout. The major elements of the launch complex shown in figure 8-1, are the vehicle assembly building (VAB), the launch control center (LCC), the mobile launcher (ML), the crawler-transporter (C-T), the crawlerway, the mobile service structure (MSS), and the launch pad.

LC-39 FACILITIES AND EQUIPMENT

VEHICLE ASSEMBLY BUILDING

The VAB is located adjacent to Kennedy Parkway, about five lmiles north of the KSC industrial area. Its purpose is to provide a protected environment for receipt and checkout of the propulsion stages and instrument unit (IU), erection of the vehicle stages and spacecraft in a vertical position on the ML, and integrated checkout of the assembled space vehicle.

The VAB, as shown in figure 8-2 is a totally enclosed structure covering eight acres of ground. It is a structural steel building approximately 525 feet high, 518 feet wide, and 716 feet long. The siding is insulated aluminum except where translucent fiberglass sandwich panels are used in part of the north and south walls.

The principal operational elements of the VAB are the low bay area and high bay area. A 92-foot wide transfer aisle extends through the length of the VAB and divides the low and high bay areas into equal segments (See figure 8-3).

Low Bay Area

The low bay area provides the facilities for receiving, uncrating, checkout, and preparation of the S-II and S-IVB stages, and the IU. The low bay area, located in the southern section of the VAB, is approximately 210 feet high, 442 feet wide, and 274 feet long. There are eight stage preparation and checkout cells, four of which are equipped with systems to simulate interface operations between the stages and the IU.

Work platforms, made up of fixed and folded sections, fit about the various sections as required. The platforms are olted, to permit vertical repositioning, to the low bay structure. Access from fixed floor levels to the work platforms is provided by stairs.

High Bay Area

The high bay area provides the facilities for erection and checkout of the S-IC stage; mating and erection operations of the S-II stage, S-IVB stage, IU, and spacecraft; and integrated checkout of the assembled space vehicle. The high bay area which is located in the northern section of the building, is approximately 525 feet high, 518 feet wide, and 442 feet long. It contains four checkout bays, each capable of accommodating a fully assembled, Saturn V space vehicle.

Access to the vehicle at various levels is provided from air conditioned work platforms that extend from either side of the bay to completely surround the launch vehicle. Each platform is composed of two biparting sections which can be positioned in the vertical plane. The floor and roof of each section conform to and surround the vehicle. Hollow seals on the floor and roof of the section provide an environmental seal between the vehicle and the platform.

Each pair of opposite checkout bays is served by a 250-ton bridge crane with a hook height of 462 feet. The wall framing between the bays and the transfer aisle is open above the 190-foot elevation to permit movement of components from the transfer aisle to their assembly position in the checkout bay.

The high bay doors provide an inverted T-shaped opening 456 feet in height. The lower portion of the opening is closed by doors which move horizontally on tracks. The upper portion of the opening is closed by seven vertically moving doors.

Utility Annex

The utility annex, located on the west side of the VAB, supports the VAB, LCC and other facilities in the VAB area. It provides air conditioning, hot water, compressed air, water for fire protection, and emergency electrical power.

Helium/Nitrogen Storage-VAB Area

The gas storage facility at the VAB provides high pressure gaseous helium and nitrogen. It is located east of the VAB and south of the crawlerway. The roof deck of the building is removable to permit installation and removal of pressure vessels through the roof. This facility is serviced from the converter/compressor facility by a 6,000 psig gaseous helium line and a 6,000 psig gaseous nitrogen line.

LAUNCH CONTROL CENTER

The LCC (figure 8-4) serves as the focal point for overall direction, control, and surveillance of space vehicle checkout and launch. The LCC is located adjacent to the VAB and at a sufficient distance from the launch pad (three miles) to permit the safe viewing of liftoff without requiring site hardening. An enclosed personnel and cabling bridge connects the VAB and LCC at the third floor level.

The LCC is a four-story structure approximately 380 by 180

LAUNCH COMPLEX 39

Figure 8-1

VEHICLE ASSEMBLY BUILDING

VEHICLE
ASSEMBLY
BUILDING

LAUNCH
CONTROL
CENTER

Figure 8-2

VAB INTERIOR

HIGH BAY
AREA

LOW BAY
AREA

Figure 8-3

LCC EXTERIOR

Figure 8-4

feet. The ground floor is devoted to service and support functions such as cafeteria, offices, shops, laboratories, the communications control room, and the complex control center. The second floor houses telemetry, RF and tracking equipment, in addition to instrumentation and data reduction facilities.

The third floor is divided into four separate but similar control areas, each containing a firing room, computer room, mission control room, test conductor platform area, visitor gallery, offices and frame rooms. Three of the four firing rooms contain control, monitoring and display equipment for automatic vehicle checkout and launch.

Direct viewing of the firing rooms and the launch area is possible from the mezzanine level through specially designed, laminated, and tinted glass windows. Electrically controlled sun louvers are positioned outside the windows.

The display rooms, offices, launch information exchange facility (LIEF) rooms, and mechanical equipment are located on the fourth floor.

The electronic equipment areas of the second and third floors have raised false floors to accommodate interconnecting cables and air conditioning ducts.

The power demands in this area are large and are supplied by two separate systems, industrial and instrumentation. The industrial power system supplies electric power for lighting, general use receptacles, and industrial units such as air conditioning, elevators, pumps and compressors. The instrumentation power system supplies power to the electronic equipment, computers, and related checkout equipment. This division between power systems is designed to protect the instrumentation power system from the adverse effects of switching transients, large cycling loads, and intermittent motor starting loads. Communication and signal cable provisions have been incorporated into the design of the facility. Cable troughs extend from the LCC via the enclosed bridge to each ML location in the VAB high bay area. The LCC is also connected by buried cableways to the ML refurbishing area and to the pad terminal connection room (PTCR) at the launch pad. Antennas on the roof

provide an RF link to the launch pads and other facilities at KSC.

MOBILE LAUNCHER

The mobile launcher (figure 8-5) is a transportable steel structure which, with the crawler-transporter, provides the capability to move the erected vehicle to the launch pad. The ML is divided into two functional areas, the launcher base and the umbilical tower. The launcher base is the platform on which a Saturn V vehicle is assembled in the vertical position, transported to a launch site, and launched. The umbilical tower, permanently erected on the base, is the means of ready access to all important levels of the vehicle during the assembly, checkout, and servicing periods prior to launch. The equipment used in the servicing, checkout, and launch is installed throughout both the base and tower sections of the ML. The intricate vehicle-to-ground interfaces are established and debugged in the convenient and protected environment of the VAB, and moved undisturbed aboard the ML to the pad.

Launcher Base

The launcher base (figure 8-6) is a two story steel structure 25 feet high, 160 feet long, and 135 feet wide. Each of the three levels provides approximately 12,000 square feet of floor space. The upper deck, designated level 0, contains, in addition to the umbilical tower, the four holddown arms and the three tail service masts. Level A, the upper of the two internal levels, contains 21 compartments and level B has 22 compartments. There is a 45-foot square opening through the ML base for first stage exhaust.

Access to the base interior is provided by personnel/equipment access doors opening into levels A and B and equipment access hatches located on levels O and A.

The base has provisions for attachment to the crawler-transporter, six launcher-to-ground mount mechanisms, and four extensible support columns.

All electrical/mechanical interfaces between vehicle systems and the VAB or the launch site are located through or

MOBILE LAUNCHER

GSCU FLOW CONTROL VALVE BOX

IU PNEUMATIC CONSOLE

IU GROUND SUPPORT
COOLING UNIT (2 UNITS)

S-IVB GAS HEAT EXCHANGER

S-IVB PNEUMATIC CONSOLE A

S-IVB PNEUMATIC CONSOLE B

S-IVB APS PNEUMATIC CONSOLE

S-II PNEUMATIC CONSOLE A&D

S-II LH2 HEAT EXCHANGER

S-II PNEUMATIC CONSOLE B

S-IC FWD UMBILICAL
SERVICE CONSOLE

S-II PNEUMATIC CONSOLE C

Figure 8-5 (Sheet 1 of 2)

MOBILE LAUNCHER

▷1 GSCU Flow Control Valve Box
Selects either GSCU for operation of one unit while the other recirculates.

▷2 Ground Support Cooling Unit
Supplies water-methanol to the heat exchanger in the IU thermal conditioning system to absorb heat in the IU generated by electronic equipment.

▷3 S-IVB Pneumatic Console A&B
Regulates and controls helium and nitrogen gases for leak testing, functional checkout, propellant loading, purge, and propellant unloading.

▷4 S-IVB APS Pneumatic Console
Regulate and distribute helium and nitrogen gases during checkout and propellant loading.

▷5 S-II LH2 Heat Exchanger A7-71
Provides gases to the S-IC stage for the following:
    1. Fuel tank pressurization
    2. LOX tank pre-pressurization
    3. Thrust Chamber jacket chilldown

▷6 S-II Pneumatic Consoles S7-41A, B, & C
Regulate, control, and monitor gases for S-II stage during standby, prelaunch, and launch.

▷7 S-IVB Gas Heat Exchanger
Supplies cold helium or hydrogen for the following:
    1. Lox and Fuel Tank Pre-Pressurization
    2. Thrust chamber jacket chilldown
    3. Pressurize engine turbine start bottle

▷8 IU Pneumatic Console
Regulates, monitors, and controls pneumatic pressure to pressurize, checkout, and test the air bearing spheres and related pneumatic and electro-mechanical circuitry.

▷9 S-IC Forward Umbilical Service Console
Supplies nitrogen from three regulation modules to S-IC stage pneumatic systems through the forward umbilical plate.

Figure 8-5 (Sheet 2 of 2)

adjacent to the base structure. A number of permanent pedestals at the launch site provide support for the interface plates and servicing lines.

The base houses such items as the computer systems test sets, digital propellant loading equipment, hydraulic test sets, propellant and pneumatic lines, air conditioning and ventilating systems, electrical power systems, and water systems. Shock-mounted floors and spring supports are provided so that critical equipment receives less than ± 0.5 g mechanically-induced vibrations. Electronic compartments within the ML base are provided with acoustical isolation to reduce the overall rocket engine noise level.

The air conditioning and ventilating system for the base provides environmental protection for the equipment during operations and standby. One packaged air conditioner provides minimal environmental conditioning and humidity control during transit. Fueling operations at the launch area require that the compartments within the structure be pressurized to a pressure of three inches of water above atmospheric pressure and that the air supply originate from a remote area free from contamination.

The primary electrical power supplied to the ML is divided into four separate services: instrumentation, industrial, in-transit and emergency. Instrumentation and industrial power systems are separate and distinct. During transit, power from the crawler-transporter is used for the water/glycol systems, computer air conditioning, threshold lighting, and obstruction lights. Emergency power for the ML is supplied by a diesel-driven generator located in the ground

facilities. It is used for obstruction lights, emergency lighting, and for one tower elevator. Water is supplied to the ML at the VAB and at the pad for fire, industrial and domestic purposes and at the refurbishment area for domestic purposes.

Umbilical Tower

The umbilical tower is an open steel structure 380 feet high which provides the support for eight umbilical service arms, one access arm, 18 work and access platforms, distribution equipment for the propellant, pneumatic, electrical and instrumentation subsystems, and other ground support equipment. The distance from the vertical centerline of the tower to the vertical centerline of the vehicle is approximately 80 feet. The distance from the nearest vertical column of the tower to the vertical centerline of the vehicle is approximately 60 feet. Two high speed elevators service 18 landings, from level A of the base to the 340-foot tower level.

The hammerhead crane is located on top of the umbilical tower. The load capacity of the crane is 25 tons with the hook extended up to 50 feet from the tower centerline. With the hook extended between 50 and 85 feet from the tower centerline, the load capacity is 10 tons. The hook can be raised or lowered at 30 feet per minute for a distance of 468 feet. The trolley speed is 110 feet per minute. The crane can rotate 360 degrees in either direction at one revolution per minute. Remote control of the crane from the ground and from each landing between levels 0 and 360 is provided by portable plug-in type control units.

MOBILE LAUNCHER LEVEL A AND B

S-IC PNEUMATIC CHECKOUT RACKS

REGULATES CONTROLS, AND MONITORS NITROGEN FOR TEST AND CHECKOUT OF PRESSURE SWITCHES IN STAGE PROPULSION SYSTEM.

SIDE 3

SIDE 2 / SIDE 4

SIDE 1
LEVEL B (LOWER)

S-IC INERT PREFILL UNIT

SUPPLIES ETHYLENE GLYCOL TO F-1 ENGINE FUEL JACKETS TO ELIMINATE ENTRAPPED AIR.

SYSTEM CHECKOUT CONSOLE

SIDE 3

SIDE 2 / SIDE 4

SIDE 1
LEVEL A (UPPER)

MOTOR CONTROL CENTER

HYDRAULIC POWER UNIT

S-IC PNEUMATIC CONSOLE
REGULATES FACILITY GASES
SUPPLIES OPERATING PRESSURES TO S-IC STAGE
AND S-IC PNEUMATIC CHECKOUT RACKS

S-IC HYDRAULIC SUPPLY AND CHECKOUT UNIT
GIMBALS FOUR OUTBOARD F-1 ENGINES
KEEPS MAIN LOX VALVES FROM FREEZING
CONTROLS F-1 ENGINE VALVES

Figure 8-6

Holddown Arms

The four holddown arms (figure 8-7) are mounted on the ML deck 90 degrees apart around the vehicle base. They position and hold the vehicle on the ML during the VAB checkout, movement to the pad and pad checkout. The arm bases have sufficient strength to support the vehicle before launch and to withstand the dynamic loads caused by engine cutoff in an abort situation. The vehicle base is held with a preloaded force of 700,000 pounds at each arm.

At engine ignition, the vehicle is restrained until proper engine thrust is achieved, at which time a signal from the countdown sequencer causes each of two identical pneumatic systems to release high pressure helium to a separator mechanism in each holddown arm. The unlatching interval for the four arms should not exceed 0.050 seconds. If any of the separators fail to operate in 0.180 seconds, release is effected by detonating an explosive nut link.

Controlled release mechanisms are used to provide a gradual release of the stage at launch, thereby keeping the dynamic loads at launch within the design capability of the vehicle. The controlled release mechanisms (figure 8-7) restrain the vehicle during the first few inches of travel. Each controlled release mechanism consists of a bracket bolted to the holddown arm base, a tapered pin fastened to the bracket, and a die linked to the vehicle. Upon holddown arm release, each tapered pin is drawn through a die as the vehicle rises through the first six inches. This reduces the diameter of the pin from its maximum to the diameter of the die. The force required to draw the pins through the die decreases linearly from maximum restraint at holddown release to zero restraint when the pins are free of the dies. The vehicle is then free with the pins remaining in the brackets and the dies traveling with the vehicle. There are provisions for as many as 16 mechanisms per vehicle. Eight mechanisms will normally be used for manned lunar missions.

Service Arms

The nine service arms provide access to the launch vehicle and support the service lines that are required to sustain the vehicle as described in figure 8-8. The service arms are designated as either preflight or inflight arms. The preflight arms are retracted and locked against the umbilical tower prior to liftoff. The inflight arms retract at vehicle liftoff, after receiving a command signal from the service arm control switches located in holddown arms 2 and 4 (see figure 8-7).

The inflight service arm launch retract sequence typically consists of the four following operations: arm unlock, umbilical carrier release, carrier withdrawal, and arm retraction and latchback. At T-15 seconds the service arms are unlocked by a signal from the terminal countdown sequencer. When the vehicle rises 3/4-inch, the primary liftoff switches on the holddown arms activate a pneumatic system which unlocks the umbilical carriers and pushes each carrier from the vehicle. If this system fails, the secondary mechanical release mechanism will be actuated when the vehicle rises approximately two inches. If both systems fail the carrier is cammed from the vehicle when it rises approximately 15 inches. Upon carrier ejection, a double pole switch activates both the carrier withdrawal and arm retraction systems. If this switch fails, it will be by-passed by a signal from the secondary liftoff switches when the vehicle rises 18 inches. Line handling devices on the S-IVB forward

and aft arms are also activated on carrier ejection. Carrier withdrawal and arm retraction is accomplished by pneumatic and/or hydraulic systems.

Tail Service Masts

The three tail service mast (TSM) assemblies, figure 8-7, support service lines to the S-IC stage and provide a means for rapid retraction at vehicle liftoff. The TSM assemblies are located on level 0 of the ML base. Each TSM is a counter-balanced structure which is penumatically/electrically controlled and hydraulically operated. Retraction of the umbilical carrier and vertical rotation of the mast is accomplished simultaneously to ensure no physical contact between the vehicle and mast. After retraction, the carrier rotates into a hood assembly which protects it from the exhaust plume.

LAUNCH PAD

The launch pad, figure 8-9, provides a stable foundation for the ML during Apollo/Saturn V launch and prelaunch operations and an interface to the ML for ML and vehicle systems. There are presently two pads at LC-39 located approximately three miles from the VAB area. Each launch site is an eight-sided polygon measuring approximately 3,000 feet across.

Launch Pad Structure

The launch pad is a cellular, reinforced concrete structure with a top elevation of 48 feet above sea level (42 feet above grade elevation). The longitudinal axis of the pad is oriented north-south, with the crawlerway and ramp approach from the south.

Located within the fill under the west side of the structure (figure 8-10) is a two-story concrete building to house environmental control and pad terminal connection equipment. On the east side of the structure, within the fill, is a one-story concrete building to house the high pressure gas storage battery. On the pad surface are elevators, staircases and interface structures to provide service to the ML and the mobile service structure (MSS). A ramp, with a five percent grade, provides access from the crawlerway. This is used by the C-T to position the ML/Saturn V and the MSS on the support pedestals. The azimuth alignment building is located on the approach ramp in the crawlerway median strip. A flame trench 58 feet wide by 450 feet long, bisects the pad. This trench opens to grade at the north end. The 700,000-pound mobile wedge-type flame deflector is mounted on rails in the trench.

An escape chute is provided to connect the ML to an underground, hardened room. This room is located in the fill area west of the support structure. This is used by astronauts and service crews in the event of a malfunction during the final phase of the countdown.

Pad Terminal Connection Room

The pad terminal connection room (PTCR) (figure 8-10) provides the terminals for communication and data link transmission connections between the ML or MSS and the launch area facilities and between the ML or MSS and the LCC. This facility also accommodates the electronic equipment that simulates the vehicle and the functions for

HOLDDOWN ARMS/TAIL SERVICE MAST

ENGINE/HOLDDOWN ARM/TAIL SERVICE MAST ORIENTATION

CONTROLLED RELEASE MECHANISMS

HOLDDOWN ARM

TAIL SERVICE MAST

Figure 8-7

MOBILE LAUNCHER SERVICE ARMS

1 S-IC Intertank (preflight). Provides lox fill and drain interfaces. Umbilical withdrawal by pneumatically driven compound parallel linkage device. Arm may be reconnected to vehicle from LCC. Retracted at T-30 seconds. Retract time is 13 seconds. Reconnect time is approximately 5 minutes.

2 S-IC Forward (preflight). Provides pneumatic, electrical, and air-conditioning interfaces. Umbilical withdrawal by pneumatic disconnect in conjunction with pneumatically driven block and tackle/lanyard device. Secondary mechanical system. Retracted at T-16.2 seconds. Retract time is 6 seconds.

3 S-II Aft (preflight). Provides access to vehicle. Arm retracted prior to liftoff as required.

4 S-II Intermediate (inflight). Provides LH$_2$ and lox transfer, vent line, pneumatic, instrument cooling, electrical, and air-conditioning interfaces. Umbilical withdrawal systems same as S-IVB Forward with addition of a pneumatic cylinder actuated lanyard system. This system operates if primary withdrawal system fails. Retract time is 6.4 seconds (max).

5 S-II Forward (inflight). Provides GH$_2$ vent, electrical, and pneumatic interfaces. Umbilical withdrawal systems same as S-IVB Forward. Retract time is 7.4 seconds (max).

6 S-IVB Aft (inflight). Provides LH$_2$ and lox transfer, electrical, pneumatic, and air-conditioning interfaces. Umbilical withdrawal systems same as S-IVB Forward. Also equipped with line handling device. Retract time is 7.7 seconds (max).

7 S-IVB Forward/IU Umbilical (inflight). For S-IVB stage, provides fuel tank vent, electrical, pneumatic, air-conditioning, and preflight conditioning interfaces. For IU, provides pneumatic, electrical, and air-conditioning interfaces. Umbilical withdrawal by pneumatic disconnect in conjunction with pneumatic/hydraulic redundant dual cylinder system. Secondary mechanical system. Arm also equipped with line handling device to protect lines during withdrawal. Retract time is 8.4 seconds (max).

8 Service Module (inflight). Provides air-conditioning, vent line, coolant, electrical, and pneumatic interfaces. Umbilical withdrawal by pneumatic/mechanical lanyard system with secondary mechanical system. Retract time is 9.0 seconds (max).

9 Command Module Access Arm (preflight). Provides access to spacecraft through environmental chamber. Arm may be retracted or extended from LCC. Retracted to 12° park position during period T-43 to T-5 minutes. Extend time is 12 seconds from this position.

Figure 8-8

LAUNCH PAD A, LC-39

Figure 8-9

LAUNCH STRUCTURE EXPLODED VIEW

Figure 8-10

checkout of the facilities during the absence of the launcher and vehicle.

The PTCR is a two-story hardened structure within the fill on the west side of the launch support structure. The launch pedestal and the deflector area are located immediately adjacent to this structure. Each of the floors of this structure measures approximately 136 feet by 56 feet. Entry is made from the west side of the launch support structure at ground level into the first floor area. Instrumentation cabling from the PTCR extends to the ML, MSS, high pressure gas storage battery area, lox facility, RP-1 facility, LH_2 facility, and azimuth alignment building. The equipment areas of this building have elevated false floors to accommodate the instrumentation and communication cables used for interconnecting instrumentation racks and terminal distributors.

The air conditioning system, located on the PTCR ground floor, provides a controlled environment for personnel and equipment. The air conditioning system is controlled remotely from the LCC when personnel are evacuated for launch. This system provides chilled water for the air handling units located in the equipment compartments of the ML. A hydraulic elevator serves the two floors and the pad level.

Industrial and instrumentation power is supplied from a nearby substation.

Environmental Control System

The ECS room located in the pad fill west of the pad structure and north of the PTCR (figure 8-10) houses the equipment which furnishes temperature and/or humidity controlled air or nitrogen for space vehicle cooling at the pad. The ECS room is 96 feet wide by 112 feet long and houses air and nitrogen handling units, liquid chillers, air compressors, a 3000-gallon water/glycol storage tank and other auxiliary electrical and mechanical equipment.

High Pressure Gas System

The high pressure gas storage facility at the pad provides the launch vehicle with high pressure helium and nitrogen. This facility is an integral part of the east portion of the launch support structure. It is entered from ground elevation on the east side of the pad. The high pressure (6,000 psig) facilities at the pad are provided for high pressure storage of 3,000 cubic feet of gaseous nitrogen and 9,000 cubic feet of gaseous helium.

Launch Pad Interface Structure

The launch pad interface structure (figure 8-11) provides mounting support pedestals for the ML and MSS, an engine access platform, and support structures for fueling, pneumatic, electric power and environmental control interfaces.

Figure 8-11

The ML at the launch pad (as well as the VAB and refurbish area) is supported by six mount mechanisms which are designed to carry vertical and horizontal loading. Four extensible columns, located near each corner of the launcher base exhaust chamber, also support the ML at the launch site. These columns are designed to prevent excessive deflections of the launcher base when the vehicle is fueled and from load reversal in case of an abort between engine ignition and vehicle liftoff.

The MSS is supported on the launch pad by four mounting mechanisms similar to those used to support the ML.

The engine servicing structure provides access to the ML deck for servicing of the S-IC engines and ML deck equipment.

Interface structures are provided on the east and west portions of the pad structure (figure 8-11) for propellant, pneumatic, power, facilities, environmental control, communications, control and instrumentation systems.

Apollo Emergency Ingress/Egress and Escape System

The Apollo emergency ingress/egress and escape system (figure 8-12) provides access to and from the Command Module (CM) plus an escape route and safe quarters for the astronauts and service personnel in the event of a serious malfunction prior to launch. Depending upon the time available, the system provides escape by either slide wire or elevator. Both means utilize the CM access arm as a component.

The slide wire egress system (figure 8-12) provides the primary means of escape. A 1 1/8-inch diameter steel cable extends from the 341.72 - foot level of the ML to a tail tower approximately 2,200 feet west of the ML. Astronauts and technicians evacuate the white room, cross the access arm, and follow a catwalk along the east and north sides of the ML to the egress platform at the 320-foot level. Here, they board the 9-man cab transporter suspended from the cable and snubbed against the egress platform. The cab is released by levers in the cab. It rides the slide wire down to the landing area where it is decelerated and stopped by an arresting gear assembly.

The secondary escape and normal egress means are the tower high speed elevators. These move between the 340 foot level of the tower and level A at 600 feet per minute. At level A, egressing personnel move through a vestibule to elevator No. 2 which takes them down to the bottom of the pad. Armored personnel carriers are available at this point to remove them from the pad area.

When the state of the emergency allows no time for retreat by motor vehicle, egressing personnel upon reaching level A of the ML slide down the escape tube into the rubber-lined blast room vestibule (figure 8-12). The escape tube consists of a short section which extends from the elevator vestibule at ML level A to side 3 of the ML base where it interfaces with a fixed portion that penetrates the pad at an elevation of 48 feet. At the lower extremity of the illuminated escape tube, a deceleration ramp is provided to reduce exit velocity, permitting safe exit for the user.

Entrance to the blast room is gained through blast-proof doors controllable from either side. The blast room floor is mounted on coil springs to reduce outside acceleration forces

to 3 to 5 g's. Twenty people may be accommodated for 24 hours. Communication facilities are provided in the room including an emergency RF link in which the receiving antenna is built into the ceiling. In the event that escape via the blast-proof doors is not possible, a hatch in the top of the blast room is accessible to rescue crews.

An underground air duct from the vicinity of the blast room to the remote air intake facility permits egress from the pad structure to the pad perimeter. Provision is made to decrease air velocity in the duct to allow personnel movement through the duct.

Emergency ingress to the CM utilizes the tower high speed elevators and the CM access arm.

Electrical Power

The electrical power for launch pad A is fed from the 69 kv main substation to switching station No. 1, where it is stepped down to 13.8 kv. The 13.8 kv power is fed to switching station No. 2 from where it is distributed to the various substations in the pad area. The output of each of the substations is 480 volts with the exception of the 4160-volt substations supplying power to the fire protection water booster pump motors and the lox pump motors.

Fuel System Facilities

The fuel facilities, located in the northeast quadrant of the pad approximately 1,450 feet from pad center, store RP-1 and liquid hydrogen.

The RP-1 facility consists of three 86,000-gallon (577,000-pound) steel storage tanks, a pump house, a circulating pump, a transfer pump, two filter-separators, an 8-inch stainless steel transfer line, RP-1 foam generating building and necessary valves, piping, and controls. Two RP-1 holding ponds, 150 feet by 250 feet with a water depth of two feet, are located north of the launch pad, one on each side of the north-south axis. The ponds retain spilled RP-1 and discharge water to drainage ditches.

The LH_2 facility consists of one 850,000-gallon spherical storage tank, a vaporizer/heat exchanger which is used to pressurize the storage tank to 65 psig, a vacuum-jacketed, 10-inch, Invar transfer line and a burn pond venting system. The internal tank pressure, maintained by circulating LH_2 from the tank through the vaporizer and back into the tank, is sufficient to provide the proper flow of LH_2 from the storage tank to the vehicle without using a transfer pump. Liquid hydrogen boiloff from the storage and ML areas is directed through vent piping to bubblecapped headers submerged in the burn pond. The hydrogen is bubbled to the surface of the 100-foot square pond where a hot wire ignition system maintains the burning process.

LOX System Facility

The lox facility is located in the northwest quadrant of the pad area, approximately 1,450 feet from the center of the pad. The facility consists of one 900,000-gallon spherical storage tank, a lox vaporizer to pressurize the storage tank, main fill and replenish pumps, a drain basin for venting and dumping of lox, and two transfer lines.

Gaseous Hydrogen Facility

This facility is located on the pad perimeter road northwest

EMERGENCY EGRESS SYSTEMS

BLAST ROOM EGRESS SYSTEM

RUBBER ROOM

BLAST ROOM

EGRESS TUNNEL
(AIR INTAKE DUCT)

1200'

FAN RM

AIR INTAKE
BUILDING

ECS

ESCAPE TUBE

ML

SLIDE WIRE EGRESS SYSTEM

EGRESS
STATION

MAIN CABLE

STATION 0

9-MAN CAB

ARRESTOR

STATION 1675

LANDING AREA

BUNKER

TAIL TOWER

WINCH

STATION 2200
STATION 2260

Figure 8-12

of the liquid hydrogen facility. The facility provides GH2 at 6,000 psig to the launch vehicle. The facility consists of four storage tanks having a total capacity of 800 cubic feet, a flatbed trailer on which are mounted liquid hydrogen tanks and a liquid-to-gas converter, a transfer line and necessary valves and piping.

Azimuth Alignment Building

The azimuth alignment building is located in the approach ramp to the launch structure in the median of the crawlerway about 700 feet from the ML positioning pedestals. The building houses the auto-collimator theodolite which senses, by a light source, the rotational output of the stable platform. A short pedestal, with a spread footing isolated from the building, provides the mounting surface for the theodolite.

Photographic Facilities

These facilities support photographic camera and closed circuit television equipment to provide real-time viewing and photographic documentation coverage. There are six camera sites in the launch pad area, each site containing an access road, five concrete camera pads, a target pole communication boxes and a power transformer with a distribution panel and power boxes. These sites cover prelaunch activities and launch operations from six different angles at a radial distance of approximately 1,300 feet from the launch vehicle. Each site has four engineering sequential cameras and one fixed, high speed, metric camera (CZR). A target pole for optical alignment of the CZR camera is located approximately 225 feet from the CZR pad and is approximately 86 feet high.

Pad Water System Facilities

The pad water system facilities supply water to the launch pad area for fire protection, cooling, and quenching. Specifically, the system furnishes water for the industrial water system, flame deflector cooling and quench, ML deck cooling and quench, ML tower fogging and service arm quench, sewage treatment plant, Firex water system, lox and fuel facilities, ML and MSS fire protection and all fire hydrants in the pad area. The water is supplied from three 6-inch wells, each 275 feet deep. The water is pumped from the wells through a desanding filter and into a 1,000,000-gallon reservoir.

Air Intake Building

This building houses fans and filters for the air supply to the PTCR, pad cellular structure and the ML base. The building is located west of the pad, adjacent to the perimeter road.

Flame Deflector

There are two flame deflectors provided at each pad; one for use and the other held in reserve. Their normal parking position is north of the launch support structure within the launch pad area. The flame deflector protects the boattail section of the Saturn V launch vehicle and the launch stand from hot gases, high pressures and flame generated by the launch vehicle during the period of engine ignition and liftoff.

MOBILE SERVICE STRUCTURE

The mobile service structure (figure 8-13) provides access to

MOBILE SERVICE STRUCTURE

Figure 8-13

those portions of the space vehicle which cannot be serviced from the ML while at the launch pad. During nonlaunch periods, the MSS is located in a parked position along side of the crawlerway, 7,000 feet from the nearest launch pad. The MSS is transported to the launch site by the C-T. It is removed from the pad a few hours prior to launch and returned to its parking area.

The MSS is approximately 402 feet high, measured from ground level, and weighs 12 million pounds. The tower structure rests on a base 135 feet by 135 feet. The top of the MSS base is 47 feet above grade. At the top, the tower is 87 feet by 113 feet.

The MSS is equipped with systems for air conditioning, electrical power, communications networks, fire protection, nitrogen pressurization, hydraulic pressure, potable water and spacecraft fueling.

The structure contains five work platforms which provide access to the space vehicle. The outboard sections of the platforms are actuated by hydraulic cylinders to open and accept the vehicle and to close around it to provide access to the launch vehicle and spacecraft. The three upper platforms are fixed but can be relocated as a unit to meet changing vehicle configurations. The uppermost platform is open, with a chain-link fence for safety. The two platforms immediately below are enclosed to provide environmental control to the spacecraft. The two lowest platforms can be adjusted vertically to serve different parts of the vehicle. Like the uppermost platform, they are open with a chain-link fence for safety.

CRAWLER-TRANSPORTER

The crawler-transporter (figure 8-14) is used to transport the mobile launcher and the mobile service structure. The ML, with the space vehicle, is transported from the vehicle assembly building to the launch pad. The MSS is transported from its parking area to and from the launch pad. After launch, the ML is taken to the refurbishment area and subsequently back to the VAB. The C-T is capable of lifting, transporting, and lowering the ML or the MSS without the aid of auxiliary equipment. The C-T supplies limited electric power to the ML and the MSS during transit.

Figure 8-14

The C-T consists of a rectangular chassis which is supported through a suspension system by four dual-tread crawler-trucks. The overall length is 131 feet and the overall width is 114 feet. The unit weighs approximately 6 million pounds. The C-T is powered by self-contained, diesel-electric generator units. Electric motors in the crawler-trucks propel the vehicle. Electric motor-driven pumps provide hydraulic power for steering and suspension control. Air conditioning and ventilation are provided where required.

The C-T can be operated with equal facility in either direction. Control cabs are located at each end and their control function depends on the direction of travel. The leading cab, in the direction of travel, will have complete control of the vehicle. The rear cab will, however, have override controls for the rear trucks only.

Maximum C-T unloaded speed is 2 mph, 1 mph with full load on level grade, and 0.5 mph with full load on a five percent grade. It has a 500-foot minimum turning radius and can position the ML or the MSS on the facility support pedestals within + two inches.

CONVERTER/COMPRESSOR FACILITY

The converter/compressor facility (CCF) converts liquid nitrogen to low pressure and high pressure gaseous nitrogen

and compresses gaseous helium to 6,000 psig. The gaseous nitrogen and helium are then supplied to the storage facilities at the launch pad and at the VAB. The CCF is located on the north side of the crawlerway, approximately at the mid-point between the VAB and the main crawlerway junction to launch pads A and B.

The facility includes a 500,000-gallon storage tank for liquid nitrogen, tank vaporizers, high pressure liquid nitrogen pump and vaporizer units, high pressure helium compressor units, helium and nitrogen gas driver/purifiers, rail and truck transfer facilities and a data link transmission cable tunnel.

The liquid nitrogen storage tank is located adjacent to the equipment building that houses the evaporators for conversion of the liquid nitrogen to high pressure gas. The liquid nitrogen is transferred to the vaporizors and compressors by pressurizing the storage tank. After vaporizing and compressing to 150 psig or 6,000 psig, the gaseous nitrogen is piped to the distribution lines supplying the VAB area (6,000 psig) and the pad (150 psig and 6,000 psig).

The gaseous helium is stored in tube-bank rail cars which are connected to the facility via a common manifold and a flexible one-inch inside diameter high pressure line. The helium passes through the CCF helium compressors which boost its pressure from the tube-bank storage pressure to 6,000 psig after which it is piped to the VAB and pad high pressure storage batteries.

Controls and displays are located in the CCF. Mass flow rates of high pressure helium, high pressure nitrogen, and low pressure nitrogen gases leaving the CCF are monitored on panels located in the CCF via cableway ducts running between the CCF and the VAB, LCC and launch pad.

ORDNANCE STORAGE AREA

The ordnance storage area serves LC-39 in the capacity of laboratory test area and storage area for ordnance items. This facility is located on the north side of the crawlerway and approximately 2,500 feet northeast of the VAB. This remote site was selected for maximum safety.

The ordnance storage installation, enclosed by a perimeter fence, is comprised of three archtype magazines, two storage buildings, one ready-storage building, an ordnance test building and a guard service building. These buildings, constructed of reinforced concrete and concrete blocks, are over-burdened where required. The facility contains approximately 10,000 square feet of environmentally controlled space. It provides for storage and maintenance of retrorockets, ullage rockets, explosive separation devices, escape rockets and destruct packages. It also includes an area to test the electro-explosive devices that are used to initiate or detonate ordnance items. A service road from this facility connects to Saturn Causeway.

VEHICLE ASSEMBLY AND CHECKOUT

The vehicle stages and the instrument unit (IU) are, upon arrival at KSC, transported to the VAB by special carriers. The S-IC stage is erected on a previously positioned mobile launcher (ML) in one of the checkout bays in the high bay area. Concurrently, the S-II and S-IVB stages and the IU are delivered to preparation and checkout cells in the low bay area for inspection, checkout, and pre-erection preparations. All components of the launch vehicle, including the Apollo spacecraft and launch escape system, are then assembled vertically on the ML in the high bay area.

Following assembly, the space vehicle is connected to the LCC via a high speed data link for integrated checkout and a simulated flight test. When checkout is completed, the crawler-transporter (C-T) picks up the ML, with the assembled space vehicle, and moves it to the launch site over the crawlerway.

At the launch site, the ML is emplaced and connected to system interfaces for final vehicle checkout and launch monitoring. The mobile service structure (MSS) is transported from its parking area by the C-T and positioned on the side of the vehicle opposite the ML. A flame deflector is moved on its track to its position beneath the blast opening of the ML to deflect the blast from the S-IC stage engines. During the prelaunch checkout, the final system checks are completed, the MSS is removed to the parking area, propellants are loaded, various items of support equipment are removed from the ML and the vehicle is readied for launch. After vehicle launch, the C-T transports the ML to the parking area near the VAB for refurbishment.

TEST SYSTEM

A computer controlled automatic checkout system is used to accomplish the VAB (high bay) and pad testing. An RCA-110A computer and the equipment necessary to service and check out the launch vehicle are installed on the ML. Also an RCA-110A computer and the display and control equipment necessary to monitor and control the service and checkout operations are installed in the LCC. The computers operate in tandem through a data link with the computer in the ML receiving commands from and transmitting data to the computer in the LCC. The physical arrangement of the LCC and the ML are illustrated in figures 8-15 and 8-6 respectively.

Test System Operation

Test system operation for Saturn V launch vehicle checkout is conducted from the firing room (see figure 8-16). During prelaunch operations, each stage is checked out utilizing the stage control and display console. Each test signal is processed through the computer complex, and is sent to the vehicle. The response signal is sent from the vehicle, through the computer complex, and the result is monitored on the display console. The basic elements of the test system and their functional relationship are shown in figure 8-17.

A switch on the control console can initiate individual operation of a system component or call up a complete test routine from the computer. The insertion of a plastic coded card key, prior to console operation, is a required precaution against improper program callup. Instructions, interruptions and requests for displays are entered into the system by keying in proper commands at the console keyboards.

A complete test routine is called up by initiating a signal at the control panel. The signal is sent to the patch distributor located in the LCC and is routed to the appropriate signal conditioning equipment where the signal is prepared for acceptance by the LCC computer complex. The LCC computer communicates with the ML computer to call up

the test routine. The ML computer complex sends the signal to the ML signal conditioning equipment and then to the stage relay rack equipment. The signal is then routed to the terminal distribution equipment and through the crossover distributor to interrogate the vehicle sensors. The sensor outputs are sent back to the ML computer complex for evaluation. The result is then sent to the LCC computer complex which routes the result to the stage console for display. Manual control of vehicle functions is provided at the control consoles. This control bypasses the computers and is sent to the vehicle by means of hardwire. The result is also sent back to the display console by hardwire.

The digital data acquisition system (DDAS) collects the vehicle and support equipment responses to test commands, formats the test data for transmission to the ML and LCC, and decommutates the data for display in the ML and LCC. Decommutated test data is also fed to the ML and LCC computers for processing and display, and for computer control of vehicle checkout. The DDAS consists of telemetry equipment, data transmission equipment and ground receiving stations to perform data commutation, data transmission and data decommutation.

The digital event evaluators are used to monitor the status of input lines and generate a time tagged printout for each detected change in input status. High speed printers in the LCC are connected to each DEE to provide a means for real time or post-test evaluation of discrete data. Two systems (DEE-3 and DEE-6) are used to monitor discrete events.

The DEE-3 is located in the PTCR with a printer located in the LCC. It monitors 768 inputs associated with propellant loading, environmental control, water control and DDAS.

The DEE-6 is located in the ML base with a printer and remote control panel in the LCC. It monitors up to 4320 discrete signals from the vehicle stage umbilicals, pad and tower ground support equipment and the DDAS.

The computer complex consists of two RCA-110A general purpose computers and peripheral equipment. This equipment includes a line printer, card reader, card punch, paper tape reader and magnetic tape transports. The peripheral equipment provides additional bulk storage for the computer, acts as an input device for loading test routines into the computer memory and as an output device to record processed data. One computer is located in the ML base and the other in the LCC. The computers are connected by underground hardwire. The LCC computer, the main control for the system accepts control inputs from test personnel at the consoles in the firing room as well as inputs from tape storage and transmits them as test commands to the ML computer. The ML computer has the test routines stored in its memory banks. These routines are called into working memory and sent as discrete signals to the launch vehicle in response to the commands received from the LCC computer. The ML computer reports test routine status, data responses and results of test to the LCC computer. It is through this link that the control equipment and personnel in the firing room are informed of the test progress.

The propellant tanking computer system (PTCS) determines and controls the quantities of fuel and oxidizer on board each stage. Optimum propellant levels are maintained and lox and LH2 are replenished as boiloff occurs during the countdown. The propellant tanking operation is monitored on the PTCS control panel.

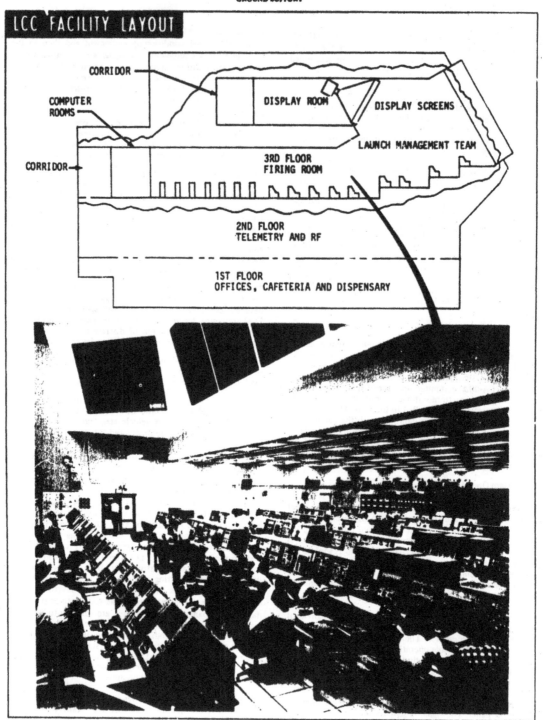

LCC FACILITY LAYOUT

CORRIDOR

COMPUTER ROOMS

CORRIDOR

DISPLAY ROOM

DISPLAY SCREENS

LAUNCH MANAGEMENT TEAM

3RD FLOOR
FIRING ROOM

2ND FLOOR
TELEMETRY AND RF

1ST FLOOR
OFFICES, CAFETERIA AND DISPENSARY

Figure 8-15

FIRING ROOM (TYPICAL)

Figure 8-16

FUNCTIONAL INTERCONNECT DIAGRAM

Figure 8-17

Visual surveillance of launch vehicle checkout is provided to the launch management team and for distribution to MSC and MSFC through the operational television system (OTV). Sixty cameras provide this capability, 27 of which are located on the ML, 15 in the pad area, 12 on the MSS and 6 in the LCC. Any camera may be requested for viewing on the 10 x 10 foot screens in the firing room.

Certain major events may be observed by members of the launch management team who occupy the first four rows in the firing room. The significant launch vehicle events which are displayed on the 10 x 10 foot screen are shown in figure 8-18.

PRELAUNCH OPERATIONS

The prelaunch operations (figure 8-19) take place in the Manned Spacecraft Operations (MSO) Building, the VAB and the launch pad.

MSO Building Activities

After receipt of the spacecraft stages at KSC, inspection, testing, assembly and integrated checkout of the spacecraft take place in the MSO building. The assembled spacecraft is transported to the VAB and mated to the launch vehicle.

VAB Activities

The VAB activities are the assembly and checkout activities which are completed in two major areas of the VAB: the high bay and the low bay.

Low Bay Activities. The low bay activities include receipt and inspection of the S-II stage, S-IVB stage and IU, and the assembly and checkout of the S-II and S-IVB stages.

The S-II stage is brought into the low bay area and positioned on the checkout dolly. An insulation leak check, J-2 engine leak check and propellant level probes electrical checks are made.

The S-IVB stage is brought into the low bay area and positioned on the checkout dolly. A fuel tank inspection, J-2 engine leak test, hydraulic system leak check and propellant level sensor electrical checks are made.

High Bay Activities. High bay activities include S-IC stage checkout, stage mating, stage systems tests, launch vehicle integrated tests, space vehicle overall tests, and a simulated flight test. High bay checkout activities are accomplished using the consoles in the firing room, the computer complex, and display equipment.

The S-IC stage is positioned and secured to the ML and access platforms are installed. The umbilicals are secured to the vehicle plates. Prepower and power-on checks are made to ensure electrical continuity. Pneumatic, fuel, lox and F-1 engine leak checks are made. Instrumentation, and range safety system checks are made.

The S-II stage is mated to the S-IC stage. The umbilicals are secured to the vehicle plates. Pre-power and power-on checks are made to ensure electrical continuity. Engine hydraulic and S-II pressurization system checks are made. Instrumentation, propulsion, propellant and range safety system checks are made.

The S-IVB stage is mated to the S-II stage and the IU is mated to the S-IVB. The S-IVB and IU umbilicals are secured to the vehicle plates. Pre-power and power-on checks are made to ensure electrical continuity. S-IVB engine hydraulic, pressurization and auxiliary propulsion system leak checks are made. S-IVB propellant, propulsion, pressurization and range safety system checks are made. IU S-band, C-band and guidance and navigation system checks are made.

Following completion of the stage system tests, launch vehicle integrated checks are accomplished. Vehicle separation, flight control, sequence malfunction and emergency detection system checks are made. The spacecraft is then mated to the launch vehicle.

After the spacecraft is mated, two space vehicle overall tests are made. Test number 1 is performed to verify RF, ordnance, pressurization, propulsion, guidance and control, propellant and emergency detection system operation. Test number 2 is performed to verify proper operation of all systems during an automatic firing sequence and flight sequence. This includes a simulated holddown arm release, electrical umbilical ejection, swing arm retraction and firing of live ordnance in test chambers. Flight type batteries are used to check out internal power.

A simulated flight test is run when the overall tests are completed. The simulated flight test verifies proper operation of the space vehicle during a normal minus count and an accelerated plus count. A normal mission profile is followed during this time. The simulated flight test ensures that the space vehicle is ready for transfer to the pad. The launch escape system is installed on the command module of the spacecraft. The ML and space vehicle are now ready for transport to the pad.

Vehicle Transfer and Pad Mating Activities

After completion of the VAB activities, the ML transports

MAJOR EVENTS

R F SILENCE	S-IVB LOX TANK PRESSURIZED	S-IC LOX TANK PRESSURIZED	COUNT	S-IVB ENGINE START
S-II PREP COMPLETE	S-IC FUEL TANK PRESSURIZED	S-IC PROPELLANTS PRESSURIZED	LIFTOFF	S-IVB CUTOFF
S-IVB PREP COMPLETE	S-IVB LH2 TANK PRESSURIZED	S-IC INTERTANK UMB DISCONNECTED	AUTOMATIC ABORT ENABLED	U/C SEPARATION
IU READY	S-IVB PROPELLANTS PRESSURIZED	S-IC FORWARD UMB DISCONNECTED	S-IC CUTOFF	PAD ABORT REQUEST
U/C READY	S-II LH2 TANK PRESSURIZED	S-II AFT UMB DISCONNECTED	S-IC/S-II SEPARATION	R F ABORT REQUEST
E D S READY	S-II PROPELLANTS PRESSURIZED	S-IC INTERTANK UMB RETRACTED	S-II ENGINE START	
RANGE SAFE	S-IC ON INTERNAL POWER	S-IC FORWARD UMB RETRACTED	S-II SECOND PLANE SEPARATION	
S-IC PREP COMPLETE	S-II ON INTERNAL POWER	S-II AFT UMB RETRACTED	LET JETTISON	
LAUNCH SEQUENCE START	S-IVB ON INTERNAL POWER	READY FOR S-IC IGNITION	S-II CUTOFF	
S-II LOX TANK PRESSURIZED	IU ON INTERNAL POWER	S-IC IGNITION	S-II/S-IVB SEPARATION	EVENT SYSTEM CALIBRATING

Figure 8-18

SCHEDULED DATES FOR SPACE VEHICLE PRELAUNCH OPERATIONS				
	AS-507		AS-508	
EVENT	START	COMPLETE	START	COMPLETE
LM Operations	24 Mar 69	23 Jun 69	28 Jun 69	26 Nov 69
Final Mate, Ascent/descent Stages		28 Apr 69		25 Jul 69
Simulated Altitude Run		6 Jun 69		4 Sep 69
Manned Altitude Run		16 Jun 69		17 Sep 69
LM/SLA Mate		23 Jun 69		26 Nov 69
CSM Operations	28 Mar 69	27 Jun 69	26 Jun 69	11 Dec 69
Docking Test		18 Apr 69		*
Combined System Tests		30 Apr 69		21 Jul 69
Unmanned Altitude Run		2 Jun 69		5 Sep 69
CSM/SLA Mate		27 Jun 69		4 Dec 69
LV VAB Low Bay Operations	24 Apr 69	20 May 69	13 Jun 69	1 Aug 69
IU Low Bay Checkout		15 May 69		1 Aug 69
S-IVB Low Bay Checkout		15 May 69		31 Jul 69
S-II Low Bay Checkout		20 May 69		16 Jul 69
LV VAB High Bay Operations	7 May 69	1 Jul 69	18 Jun 69	12 Dec 69
S-IC Erection		7 May 69		18 Jun 69
LV Erection		22 May 69		1 Aug 69
Stage Power ON		28 May 69		7 Aug 69
LV Malfunction Overall Test		12 Jun 69		*
LV Service Arm Overall Test		19 Jun 69		5 Dec 69
Spacecraft Erection		1 Jul 69		12 Dec 69
Space Vehicle VAB Operations	1 Jul 69	8 Sep 69	12 Dec 69	15 Dec 69
Transfer to Pad	8 Sep 69	8 Sep 69	15 Dec 69	15 Dec 69
Pad Operations	8 Sep 69	14 Nov 69	15 Dec 69	9 Mar 70
LV Power ON		10 Sep 69		17 Dec 69
Service Arm Pressure Test		16 Sep 69		*
LV Flight Systems Redundancy Test		25 Sep 69		16 Jan 70
Space Vehicle Flight Electrical Mating		25 Sep 69		16 Jan 70
Space Vehicle Back-up Guidance Test		29 Sep 69		20 Jan 70
Space Vehicle Flight Readiness Test		30 Sep 69		21 Jan 70
Space Vehicle Hypergolic Loading		18 Oct 69		10 Feb 70
S-IC RP-1 Loading		20 Oct 69		11 Feb 70
CDDT - Wet/Dry		29 Oct 69		20 Feb 70
Space Vehicle Inspection and Workoff		7 Nov 69		2 Mar 70
Countdown	8 Nov 69	14 Nov 69	3 Mar 70	9 Mar 70

* Dates not available

Figure 8-19

the assembled space vehicle to the launch pad. Approximately eight hours are required for this operation. The space vehicle/ML are then interfaced with the launch pad.

Pad Activities

In general, once the vehicle and ML have been mated to the pad facility, two major operations must be performed. The first is to verify the readiness of the launch vehicle, spacecraft and launch facility to perform the launch sequence and the second is to complete the launch operation.

The Countdown Demonstration Test (CDDT) verifies that the launch vehicle and the ground support vehicle are in launch status. The CDDT is performed in two phases, the wet CDDT and the dry CDDT.

The wet CDDT is performed the same as the launch countdown with the following major exceptions:

1. Service Arms No. 1 and 2 are not retracted

2. Digital Range Safety Command System test code plugs are used instead of flight code plugs

3. Hypergol cartridges, igniters, initiators, safe and arm devices, and exploding bridgewire detonators are inert

4. Astronauts do not board the spacecraft.

The dry CDDT is performed the same as the last 3 1/2 hours of the launch countdown with the following major exceptions:

1. Launch vehicle cryogenic propellants are not on board

2. Service Arms No. 1 and 2 are not retracted

3. The primary damper is not disconnected

4. Service Arm No. 9 is reconnected as soon as the system has stabilized in the park position

5. Hypergol cartridges, igniters, initiators, safe and arm devices, and exploding bridgewire detonators are inert.

Following the CDDT, preparations for the actual countdown are started. The preparations include items which would either compromise the safety of the vehicle if done later in the countdown, or impose additional constraints on pad access during the final phases of the countdown.

Approximately four days before the launch readiness day, the countdown begins and the space vehicle is subjected to the final checkout and servicing operations required for launch.

The final phase of the countdown starts approximately nine hours prior to liftoff. During the final phase, the cryogenics are loaded, conditioned and pressurized. Final checks are made on all subsystems. The propulsion systems are serviced and prepared for launch. All onboard spheres are brought up to flight pressure and the crew mans the Command Module. Figure 8-20 lists the significant events of this period.

By the time spacecraft closeout is complete, most major operations have been completed. Propellants are being replenished as required to supplement cryogens lost due to boiloff. Boiloff will continue until the various stage vent valves are closed for tank prepressurization and some vapor may be noticeable.

With the start of the automatic sequence at T-187 seconds, the final operations required for launch begin. All pneumatic and propellant supply lines are vented and purged to prevent damage to the vehicle at umbilical release. The vehicle is switched to internal power, necessary purges are put in launch mode and some service arms are retracted.

At T-8.9 seconds, the S-IC ignition command is given. This signal starts engine no. 5 and causes the venting of the remaining high pressure pneumatic lines.

At T-0 seconds, the launch commit signal is given, causing the holddown arms to retract hydraulically. These four arms restrain the launch vehicle until a satisfactory thrust level is achieved after which the controlled release assemblies provide for gradual release of the vehicle during liftoff.

HOLD AND RECYCLE CRITERIA

Interruption of the countdown due to equipment failure, weather or other causes may occur at any time. When the countdown is interrupted, subsequent feasible actions depend on the function taking place at the time. These actions include holding and/or recycling. Feasible actions also, in some cases, are affected by previous operations conducted on the vehicle, such as the number of pressure cycles the propellant tanks have undergone.

A hold is defined as an interruption of the countdown for unfavorable weather, repair of hardware, or correction of conditions unsatisfactory for launch or flight.

In a recycle, the countdown is stopped and returned to a designated point as specified in the launch mission rules.

For a scrub, the launch attempt must be rescheduled for a later window.

A turnaround comprises the actions required to recycle, hold until the countdown can be resumed for a specific window, and complete the countdown from the re-entry point to T-0.

Decision/repair time is the time available to make decisions and/or conduct repair operations before initiating the count.

Launch Constraints

Various operational, launch vehicle, spacecraft and support equipment factors affect hold/recycle processing actions. Several of these factors are briefly discussed in the following paragraphs.

The length of the launch window on any launch date (four to five hours for lunar missions) is a mission peculiar constraint. This constraint determines the maximum hold limit for the countdown period between T-8 hours 15 minutes (start of cryogenic loading) and T-22 minutes (commence S-II stage start bottle chilldown).

Launch vehicle batteries have a life of 120 hours following activation. The batteries are installed in the vehicle at T-27.5 hours and are activated 28.5 hours prior to their installation. Assuming a six hour mission requirement for the batteries, and a countdown programmed hold of six hours, the battery life expended in a normal count is 68 hours, leaving an available battery life of 52 hours.

The Safe and Arm (S&A) devices are remotely controlled ordnance items used to make safe and/or arm the launch vehicle propellant dispersion systems. The devices are certified at T-109 hours. Recertification is required in seven calendar days. Device removal from the vehicle is required to perform recertification. The devices are installed at T-102 hours and connected at T-11.5 hours. In a normal count, the allowable S&A device life is 59 hours.

The S-IC Forward Service Arm provides the environmental conditioning interface between ground support facilities and the S-IC/S-II interstage area. Launch mission rules specify a recycle to T-22 minutes if the countdown is interrupted between service arm retraction and S-IC ignition. At T-22 minutes, the decision to hold or scrub is made.

The S-IVB Auxiliary Propulsion System (APS) module is serviced with hyperbolic propellants. Pressurization of the system is done at T-40 minutes. Depressurization and gas removal must be accomplished in the event of a scrub if access is required in the S-IVB access control area. This task takes two hours and requires use of the Mobile Service Structure (MSS).

The Command and Service Module (CSM) fuel cell cryogenics provide the electrical power for the spacecraft. The water resulting from the reaction in the fuel cell is used for drinking purposes during space missions. The cryogenic tanks are

TYPICAL PRELAUNCH SEQUENCE (SELECTED EVENTS FROM T-9:30:00.0 TO LIFTOFF)					
TIME BEFORE FIRST MOTION (HR:MIN:SEC)			TIME BEFORE FIRST MOTION (HR:MIN:SEC)		
START	STOP	EVENT	START	STOP	EVENT
9:30:00.0	LIFTOFF	Switch LV Environmental Conditioning From Air To GN$_2$	1:40:00.0	1:10:00.0	EDS Test
8:59:00.0	1:30:00.0	G&C System Checks	1:00:00.0	0:35:00.0	S-IC Fuel Level Adjust
8:59:00.0	LIFTOFF	ST-124M3 GN$_2$ Sphere Pressurization	1:00:00.0	0:00:08.9	S-IVB Ambient GHe Supply Bottle Pressurization
8:59:00.0	1:00:00.0	S-IVB Ambient GHe Supply Bottle Prepressurization	0:57:00.0	0:47:00.0	Azimuth Laying
8:59:00.0	8:54:00.0	LVDC Sector Sum Check	0:51:30.0	LIFTOFF	Final CCS Checks
8:57:00.0	6:27:00.0	S-IC Lox Tank Purge (GN$_2$)	0:51:25.0	0:46:25.0	Preflight Command System Test
8:34:00.0	7:39:00.0	S-II Lox Tank Prepressurization (GN$_2$)	0:45:00.0	LIFTOFF	Final Tracking System Check
8:34:00.0	7:42:00.0	S-II LH$_2$ Tank Purge (GHe)	0:43:00.0	0:05:00.0	CM Access Arm in Park Position
8:34:00.0	8:16:00.0	S-II LH$_2$ Recirculation Line Purge (GHe)	0:42:00.0	LIFTOFF	LES Armed
8:15:00.0	8:00:00.0	S-IVB Lox Chilldown Pump Cavity Purge	0:42:00.0	LIFTOFF	GNC System Checks
8:15:00.0	LIFTOFF	ST-124M3 System ON	0:42:00.0	0:37:00.0	IU Gimbal Checks
8:15:00.0		Start LV Cryogenic Loading	0:40:00.0	0:32:00.0	Range Safety Command Checks (Closed Loop)
8:00:00.0	7:55:00.0	Prepare to Launch Test	0:34:30.0	0:00:08.9	S-IVB Engine GHe Bottle Pressurization
8:00:00.0	LIFTOFF	IU Electrical Disconnect Purge	0:31:30.0	0:30:00.0	S-II LH$_2$ Chilldown Pump Test
8:00:00.0	7:40:00.0	S-II Turbine Start Bottle Purge (GHe)	0:31:30.0	0:30:00.0	S-IVB Lox/LH$_2$ Chilldown Pump Test
8:00:00.0	0:04:00.0	S-II Auxiliary Hydraulic Pump ON	0:30:00.0	0:26:00.0	Power Transfer Test
8:00:00.0	3:28:00.0	Operate DEE-3 System	0:25:00.0	LIFTOFF	Range Safety System ON
7:43:00.0	7:28:00.0	S-IVB Lox Tank Purge (GHe)	0:22:00.0	0:04:37.0	S-II Turbine Start Bottle Chilldown (GH$_2$)
7:42:00.0	4:54:00.0	S-II LH$_2$ Tank Preconditioning	0:20:30.0	LIFTOFF	Final SV Telemetry System Checks
7:40:00.0	7:28:00.0	S-II Engine Turbopump Purge	0:20:00.0	0:14:30.0	S-IVB Turbine Start Bottle Purge
7:39:00.0	7:04:00.0	S-II Lox Tank Purge (GN$_2$)	0:15:00.0	0:10:00.0	S-IVB Engine Thrust Chamber Purge
7:39:00.0	7:28:00.0	S-IVB J-2 Engine Turbopump Purge	0:14:30.0	0:05:30.0	S-IVB Turbine Start Bottle Chilldown
7:31:00.0	4:11:00.0	S-IVB LH$_2$ Tank Purge (GHe)	0:13:00.0	0:08:00.0	S-II Engine Thrust Chamber Purge (GHe)
7:28:00.0	7:13:00.0	S-II Engine Thrust Chamber Purge (GHe)	0:10:00.0	LIFTOFF	Prepare to Launch Test
7:28:00.0	7:24:00.0	S-IVB Lox Tank Precool(0-5%)	0:10:00.0	0:06:00.0	S-IC Fuel Jacket Topping (Glycol)
7:24:00.0	7:07:00.0	S-IVB Lox Tank Fast Fill (5-96%)	0:10:00.0	0:00:08.9	S-IVB Engine Thrust Chamber Chilldown
7:07:00.0	7:04:00.0	S-IVB Lox Tank Slow Fill (96-99%)	0:08:00.0	0:00:08.9	S-II Engine Thrust Chamber Chilldown
7:05:00.0	LIFTOFF	S-IC Engine Thermal Conditioning	0:05:30.0	LIFTOFF	S&A Devices Armed
7:05:00.0	6:30:00.0	IU CCS Checks	0:05:30.0	0:05:00.0	S-IVB Turbine Start Bottle Pressurization
7:04:00.0	6:54:00.0	S-II Lox Tank Precool (0-5%)	0:05:00.0	0:04:00.0	Remove Q-ball Cover
6:54:00.0	6:44:00.0	S-II Lox Tank Fast Fill (5-40%)	0:05:00.0	0:04:50.0	Retract and Lock CM Access Arm
6:44:00.0	6:30:00.0	S-II Lox Tank Fast Fill (40-96%)	0:04:37.0	0:03:07.0	S-II Turbine Start Bottle Pressurization
6:38:00.0	6:30:00.0	Preflight Command System Test, Houston	0:04:30.0	LIFTOFF	Terminal Countdown Sequencer Armed
6:30:00.0	6:27:00.0	S-II Lox Tank Slow Fill (96-99%)	0:03:07.0		Start Automatic Sequence
6:27:00.0	6:02:00.0	S-IC Lox Precool	0:03:07.0	0:00:30.0	S-II Lox Tank Prepressurization
6:02:00.0	5:45:00.0	S-IC Lox Tank Slow Fill (0-6.5%)	0:02:47.0	0:00:08.9	S-IVB Lox Tank Prepressurization
5:45:00.0	4:57:00.0	S-IC Lox Tank Fast Fill (6.5-98%)	0:01:37.0	0:00:04.1	S-IC Fuel Tank Prepressurization
5:45:00.0	0:03:06.0	S-II Lox Tank Replenish	0:01:37.0	0:00:30.0	S-II LH$_2$ Tank Prepressurization
5:45:00.0	0:03:06.0	S-IC Lox Tank Replenish	0:01:37.0	0:00:08.9	S-IVB LH$_2$ Tank Prepressurization
4:57:00.0	0:00:08.9	S-IC GHe Supply Bottle Pressurization	0:01:12.0	LIFTOFF	S-IC Lox Tank Prepressurization
4:57:00.0	4:55:00.0	S-IC Lox Tank Slow Fill (98-100%)	0:00:50.0		Transfer to Internal Power
4:55:00.0	0:03:06.0	S-IC Lox Tank Replenish	0:00:40.0	LIFTOFF	S-IVB LH$_2$ Vent Directional Control Flight Position ON
4:54:00.0	4:36:00.0	S-II LH$_2$ Precool (0-5%)	0:00:30.0	0:00:17.0	Retract and Lock S-IC Intertank Service Arm
4:36:00.0	4:16:00.0	S-II LH$_2$ Fast Fill (5-98%)	0:00:30.0	LIFTOFF	S-IC Engine Hydraulic System Flight Activation
4:16:00.0	4:11:00.0	S-II LH$_2$ Slow Fill (98-100%)	0:00:17.0	LIFTOFF	Guidance Reference Release
4:11:00.0	LIFTOFF	Q-Ball Power and Heaters ON	0:00:16.2	0:00:11.0	Retract and Lock S-IC Forward Service Arm
4:11:00.0	0:00:08.9	S-IVB Lox Tank GHe Supply Bottle Pressurization	0:00:08.9	0:00:06.7	S-IC Ignition Command
4:11:00.0	3:58:00.0	S-IVB LH$_2$ Precool (0-5%)	0:00:05.3	LIFTOFF	Monitor S-IC Engine Thrust Buildup
3:58:00.0	3:33:00.0	S-IVB LH$_2$ Fast Fill (5-98%)	0:00:01.0	LIFTOFF	Monitor Thrust OK All S-IC Engines
3:33:00.0	3:28:00.0	S-IVB LH$_2$ Slow Fill (98-100%)	0:00:01.0	0:00:00.0	Holddown Arm Release
3:33:00.0	0:03:06.0	S-II LH$_2$ Replenish			
3:28:00.0	0:03:06.0	S-IVB LH$_2$ Replenish			
3:28:00.0	0:52:30.0	Astronaut Loading			

Figure 8-20

loaded to sufficient capacity to tolerate a 56 hour delay for a normal count. Water generated by fuel cell operation must be drained if a hold will exceed 17 hours. Cryogenic replenishment is normally required if turnaround exceeds 56 hours. The MSS is required for CSM cryogenic servicing.

The Lunar Module (LM) supercritical helium (SHe) system provides pressurization for the LM descent engine. After loading and topping, the pressure within the SHe tank can increase at a rate up to 10 psi per hour. The maximum allowable tank pressure, the rate of tank pressure buildup and the elapsed time from SHe loading to engine ignition significantly affect the hold constraint imposed by the SHe system.

Capacities of launch support facilities and equipment such as the gaseous hydrogen and helium facilities, the cryogenic storage facilities and the ground hydraulic supply unit affect hold and recycle capability. For example, the gaseous hydrogen system can support four complete cycles of start tank chilldown operations from T-22 minutes to T-4.5 minutes before recharging of the storage battery is required. Recharging is accomplished by mobile units which cannot be moved into position until launch vehicle cryogenics are downloaded. A second example, maximum hold times are affected by capacity of the lox and LH_2 storage facilities. Cryogenic boiloff and losses incurred when recycle operations require downloading of propellants may require replenishment of the storage tanks to support a subsequent launch attempt.

Turnaround Requirements

Vehicle turnaround operations are required if a countdown interruption occurs after the initiation of vehicle cryogenic loading and if vehicle access is necessary to correct the problem causing the interruption. Turnaround operations can also be required if it becomes apparent that the maximum hold limits will be exceeded as a result of the countdown interruption.

For the functions necessary to accomplish vehicle turnaround during the countdown period from the start of cryogenic loading to liftoff, four basic turnaround alternatives are considered. A brief description of each of these alternatives is provided in the following paragraphs.

In the first turnaround alternative the MSS is not required for vehicle access. Inspections, systems checks and/or repairs may be conducted within the space vehicle. Required turnaround operations to permit vehicle repairs include cryogen downloading, tank purge, hazardous gas checks, installation of access platforms and disconnecting S&A devices. When repairs are completed, the S&A devices are reconnected, access platforms are removed and the countdown is resumed at T-9 hours. This provides a period for pad clearing prior to start of cryogenic loading operations at T-8 hours 15 minutes.

In the second turnaround alternative, the necessary inspections, systems checks and/or repairs require vehicle access not obtainable from the mobile launcher. Movement and positioning of the MSS is accomplished in addition to the turnaround operations required for the first alternative. The duration of available decision/repair time varies within the time frame in which the countdown interrupt occurs and the waivers exercised by the launch managers. The countdown is resumed at T-9 hours.

The third turnaround alternative is required if LM SHe servicing is necessary. The MSS is required for this alternative. The countdown is resumed at T-17 hours, the point at which SHe topoff is completed.

The fourth turnaround alternative is required when it becomes apparent that correction of the problem causing the countdown interruption will exceed more than one maximum space vehicle system hold constraint. The operations associated with this alternative permits replacement of the launch vehicle and spacecraft batteries, reservicing of the spacecraft cryogenic systems and reservicing of the LM SHe system. The decision/repair time available is dependent upon the time frame in which the interruption occurs and the selection of the subsequent launch window. The countdown is resumed at T-28 hours, the point at which the pad is cleared for the installation of launch vehicle batteries and the disconnec and removal of spacecraft cryogenic system servicing equipment.

LAUNCH CUTOFF INTERLOCK

The Saturn V vehicle countdown is automatically controlled from T-187 seconds to T+6 seconds by the terminal countdown sequencer (TCS), located in the mobile launcher (ML). This portion of the countdown is referred to as the automatic sequence.

Three interlocks must be activated prior to T-187 seconds to enable sequencer start: the TCS must be manually armed; the firing command preparation complete summation interlock must be activated; and the firing command enable switch on the S-IC launch control panel must be depressed. With these three interlocks activated, the T-187 second pulse from the countclock will start the TCS. The firing command preparation complete summation interlock includes many critical system parameters, events, and conditions, e.g., gas generator valves closed, all main fuel valves closed, EDS ready, exhaust igniters installed, ordnance preparation complete, spacecraft ready for launch, hydraulic pressure OK, ignition source voltage OK, launch support preparation complete, all checkout valves in ground return position, IU ready for firing, and many similar items. Improper operation or failure of any of these items will inhibit activation of the firing command preparation complete summation interlock and consequently prevent the start of the TCS.

Once the automatic sequence has been initiated, it can be stopped only by a cutoff signal; there are no provisions for holding.

Manual Cutoff Capability

Manual closure of the cutoff switch directly initiates a cutoff command. This capability may be used if the test conductor observes a condition serious enough to warrant cutoff. Manual cutoff is available at any time from TCS start to umbilical separation.

Instant Cutoff Interlocks

The TCS may be terminated automatically by any one of the following six cutoff interlocks:

1. S-IC stage logic cutoff. The cutoff signal is caused by (a) the presence of an engine shutdown signal on the stop solenoids from internal sources; or (b) a malfunction in either of the redundant range safety command destruct systems. The range safety command receiver system check is enabled from T-17 seconds to T-50 milliseconds (ms).

2. S-IC main fuel valve failure. The cutoff signal is caused by (a) both main fuel valves on any engine open prior to hypergol rupture or (b) on any engine, one main fuel valve open while the other valve remains closed (to prevent lox-rich condition in the engines). For main fuel valve failure, engines will shut down in a 3-2 sequence with the first three engines shutting down at cutoff and the other two 100 ms later. If the failure occurs in engine 2 or 4, a preferred engine shutdown sequence will occur, i.e., 2, 4, and 5 followed by 1 and 3. If the failure occurs in engine 1, 3, or 5, the normal engine shutdown sequence will occur, i.e., 1, 3, and 5 followed by 2 and 4. If cutoff is caused by other than main fuel valve failure, the normal shutdown sequence will occur.

3. Sequencer power supply failure. The cutoff signal is caused by an out-of-tolerance value of the voltage supply. A new TCS under development having a battery backup will eventually eliminate this interlock.

4. S-IC voltage failure. The cutoff signal is caused by improper voltage output from either the stage main bus (+ 1D11) or the stage instrumentation bus (+1D21).

5. Emergency detection system (EDS) failure. The EDS failure interlock is enabled from T-8.9 seconds to T-50 ms. A cutoff signal is caused by one of the three manual cutoff commands from the spacecraft or loss of one of the three EDS voting logic buses.

6. IU failure cutoff. This interlock is enabled from T-8.9 seconds to T-50 ms. During this period, a loss of IU ready to launch will initiate cutoff. The IU ready to launch interlock monitors the IU power systems, the flight computer, and the presence of the S-IC ignition command.

MISSION CONTROL MONITORING

TABLE OF CONTENTS

INTRODUCTION

Mission control monitoring provides, at the various operational levels, the information required to control, direct and evaluate the mission from prelaunch checkout through recovery. The monitoring function during vehicle flight includes space vehicle tracking, receipt and evaluation of flight crew and launch vehicle status, transmittal of up-data commands to the onboard computer and voice communications with the flight crew. The facilities used in the accomplishment of the monitoring function include an assembly, checkout and launch facility, a central flight control facility, a worldwide network of monitoring stations and a real-time display system.

Associated with the flight crew in mission control operations are the following organizations and facilities:

1. Mission Control Center (MCC), Manned Spacecraft Center, Houston, Texas. The MCC contains the communication, computer, display, and command systems to enable the flight controllers to effectively monitor and control the space vehicle.

2. Kennedy Space Center, Cape Kennedy, Florida. The space vehicle is assembled and launched from this facility. Prelaunch, launch, and powered flight data are collected by the Central Instrumentation Facility (CIF) at KSC from the launch pads, CIF receivers, Merritt Island Launch Area (MILA), and the downrange Air Force Eastern Test Range (AFETR) stations. This data is transmitted to MCC via the Apollo Launch Data System (ALDS). Also located at KSC is the Impact Predictor (IP).

3. Goddard Space Flight Center (GSFC), Greenbelt, Maryland. GSFC manages and operates the Manned Space Flight Network (MSFN) and the NASA Communications (NASCOM) networks. During flight, the MSFN is under operational control of the MCC.

4. George C. Marshall Space Flight Center (MSFC), Huntsville, Alabama. MSFC, by means of the Launch Information Exchange Facility (LIEF) and the Huntsville Operations Support Center (HOSC), provides launch vehicle systems real-time support to KSC and MCC for preflight, launch, and flight operations.

A block diagram of the basic flight control interfaces is shown in figure 9-1.

VEHICLE FLIGHT CONTROL CAPABILITY

Flight operations are controlled from the MCC. The MCC is staffed by flight control personnel who are trained and oriented on one program and mission at a time. The flight control team members perform mission planning functions and monitor flight preparations during preflight periods. Each member becomes and operates as a specialist on some aspect of the mission.

MCC ORGANIZATION

The MCC has two control rooms for flight control of manned space flight missions. Each control room, called a Mission Operations Control Room (MOCR), is used independently of the other and is capable of controlling individual missions. The control of one mission involves one MOCR and a designated team of flight controllers. Staff Support Rooms (SSR's), located adjacent to the MOCR, are manned by flight control specialists who provide detailed support to the MOCR. Figure 9-2 outlines the organization of the MCC for flight control and briefly describes key responsibilities. Information flow within the MOCR is shown in figure 9-3.

The consoles within the MOCR and SSR's permit the necessary interface between the flight controllers and the spacecraft. The displays and controls on these consoles and other group displays provide the flight controllers with the capability to monitor and evaluate data concerning tthe mission.

Problems concerning crew safety and mission success are identified to flight control personnel in the following ways:

1. Flight crew observations.

2. Flight controller real-time observations.

3. Review of telemetry data received from tape recorder playback.

4. Trend analysis of actual and predicted values.

5. Review of collected data by systems specialists.

6. Correlation and comparison with previous mission data.

7. Analysis of recorded data from launch complex testing.

The facilities at the MCC include an input/output processor designated as the Command, Communications and Telemetry System (CCATS) and a computational facility, the Real-Time Computer Complex (RTCC). Figure 9-4 shows the MCC functional configuration.

The CCATS consists of three Univac 494 general purpose computers. Two of the computers are configured so that either may handle all of the input/output communications for two complete missions. One of the computers acts as a dynamic standby. The third computer is used for nonmission activities.

The RTCC is a group of five IBM 360 large scale, general purpose computers. Any of the five computers may be designated as the mission operations computer (MOC). The MOC performs all the required computations and display formatting for a mission. One of the remaining computers will be a dynamic standby. Another pair of computers may be used for a second mission or simulation.

SPACE VEHICLE TRACKING

From liftoff of the launch vehicle to insertion into orbit, accurate position data are required to allow the Impact Predictor (IP) and the RTCC to compute a trajectory and an orbit. These computations are required by the flight controllers to evaluate the trajectory, the orbit, and/or any abnormal situations to ensure safe recovery of the astronauts. The launch tracking data are transmitted from the AFETR sites to the IP and then to the RTCC via high-speed data communications circuits at the rate of ten samples per second (s/s). The IP also generates a state vector smooth sample which is transmitted to the RTCC at a rate of two s/s. (A state vector is defined as spacecraft inertial position and inertial rate of motion at an instant of time.) The message from the IP to the RTCC alternately contains one smoothed vector, then five samples of best radar data. Low speed tracking data are also transmitted via teletype (TTY) to MCC

at a rate of one sample per six seconds from all stations actively tracking the spacecraft. Figure 9-5 shows data flow from liftoff to orbital insertion.

As the launch vehicle is boosting the spacecraft to an altitude and velocity that will allow the spacecraft to attain earth orbit, the trajectory is calculated and displayed on consoles and plotboards in the MOCR and SSR's. Also displayed are telemetry data concerning status of launch vehicle and spacecraft systems. If the space vehicle deviates excessively from the nominal flight path, or if any critical vehicle condition exceeds tolerance limits, or if the safety of the astronauts or range personnel is endangered, a decision is made to abort the mission.

During the orbit phase of a mission, all stations that are actively tracking the spacecraft will transmit the tracking data through GSFC to the RTCC by teletype, at a frequency of one sample every six seconds. If a thrusting maneuver is performed by the spacecraft, high-speed tracking data at the rate of five s/s is transmitted in addition to the teletype data.

Any major maneuver during a mission is planned to occur during or just prior to acquisition by a tracking station that can relay high-speed tracking data to the MCC. This is to ensure that data is available for the calculation of the new spacecraft orbit and ephemeris.

Approximately 25 minutes prior to anticipated spacecraft acquisition by a tracking station, a message giving time, antenna position coordinates, and range is dispatched to that station. This information is computed from the ephemeris and is used by station personnel to pre-position the antenna and enable spacecraft acquisition with minimum delay.

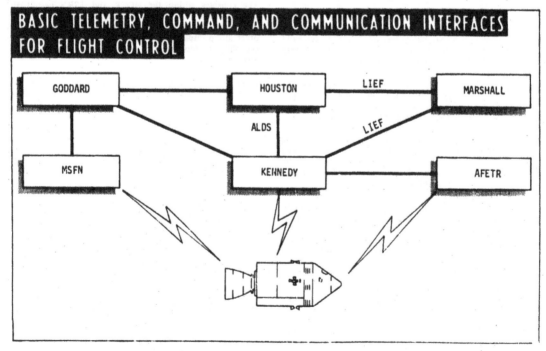

Figure 9-1

MOCR AND SSR ORGANIZATION

Figure 9-2

Figure 9-3

Figure 9-4

COMMAND SYSTEM

The Apollo ground command systems have been designed to work closely with the telemetry and trajectory systems to provide a method of "closed-loop" command which includes the astronauts and flight controllers as vital links in the commanding operation. For example, analysis of spacecraft data by flight controllers results in a command to alter an observed condition. The effects of the command will be observed in subsequent data presented to the flight control team. This verifies the correct execution of the command and closes the loop. In some cases, such as maneuvering the spacecraft, the command may not be immediately executed, but instead, the astronaut will maneuver the spacecraft at the optimum time specified by the command. The resulting telemetry and trajectory will reflect the maneuver and close the commanding loop.

To prevent spurious commands from reaching the space vehicle, switches on the Command Module console block uplink data from the onboard computers. At the appropriate times, the flight crew will move the switches from the BLOCK to ACCEPT position and thus permit the flow of uplink data.

With a few exceptions, commands to the space vehicle fall into two categories, real-time commands and command loads (also called computer loads, computer update, loads, or update). Among the exceptions is the "clock word" command. This command is addressed to the onboard timing system and is used when the downlink (telemetry) time word and the ground timing system are out of tolerance.

Real-Time Commands

Real-time commands are used to control space vehicle systems or subsystems from the ground. The execution of a real-time command results in immediate reaction by the affected system. Real-time commands correspond to unique space vehicle hardware and therefore require careful pre-mission planning to yield commands which provide alternate systems operation in the event of an anticipated failure. Pre-mission planning also includes commands necessary to initiate mission contingency plans. Descriptions of several real-time commands used by the Booster System Engineer follow.

The ALTERNATE SEQUENCE real-time commands permit the onboard Launch Vehicle Digital Computer (LVDC) to deviate from its normal program and enter a predefined, alternate sequence of program steps. For example, should the S-II engines fail to ignite or the engines shut down prematurely, an alternate switch selector sequence would jettison the S-II stage and the S-IVB stage would be used to attain a parking orbit.

The SEQUENCE INHIBIT real-time command provides the capability to inhibit a programed sequence, usually a maneuver. Each sequence must be separately inhibited with the command being processed immediately after LVDC acceptance. The maneuvers may be inhibited in any random order required during the mission. If an update for a particular command is received after the inhibit for that command, the inhibit is removed and the maneuver will occur at the update time specified.

Other examples of real-time commands are: LH_2 VENT CLOSED, LH_2 VENT OPEN, LOX VENT OPEN, TERMINATE, SET ANTENNA OMNI, SET ANTENNA LO-GAIN, and SET ANTENNA HI-GAIN.

Real-time commands are stored prior to the mission in the Command Data Processor (CDP) at the applicable command site. The CDP, a Univac 642B general purpose digital computer, is programed to format, encode, and output commands when a request for uplink is generated.

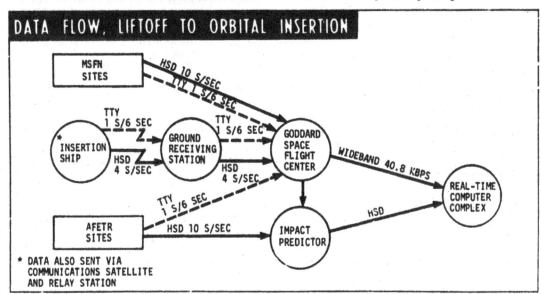

Figure 9-5

Command Loads

Command loads are generated by the real-time computer complex on request of flight controllers. Command loads are based on the latest available telemetry and/or trajectory data. Due to the nature of these commands, the data structure cannot be determined prior to the mission but must be calculated as a result of real-time data. A command load, for example, may define the exact conditions under which a thrust may be applied that will change a faulty orbit to the desired orbit.

The RTCC operating personnel take data supplied by the flight controllers requesting the command load, and by selecting the appropriate computer program, cause the computer to "make up" a command load. When the load is "ready", it is reviewed by the responsible flight controller via the display system. When the load is approved, it is transferred via NASCOM, in the form of high-speed data and/or teletype messages to the appropriate site and stored in the command data processor. The CCATS will retain the load in memory where it is available for retransmission should difficulties be encountered in the transfer procedure. When the command load is properly stored in the site's command data processor, a load validation message is sent to the CCATS and to the flight controller.

Flight controllers typically required to generate a command load include the Booster Systems Engineer (BSE), the Flight Dynamics Officer (FDO), the Guidance Officer (GUIDO), and the Retrofire Officer (RETRO).

Prior to the acquisition of the space vehicle by a site, the flight controllers requiring command capability during the pass indicate their requirements to the Real-Time Command Controller (R/T CMD) in the CCATS area. The R/T CMD enables the circuitry to permit the command function for console/site combinations requested.

When the space vehicle has been acquired by the site, it will be announced over one of the voice coordination loops and each flight controller will execute his commands according to the priority assigned by real-time decision. Disposition of each command is displayed by indicator lights on the command panel. These indicator lights are operated by CCATS in response to the verification and/or reject messages received from the site. Typical command loads (BSE) are described below.

The SECTOR DUMP command causes the LVDC to telemeter the entire contents of one memory sector, or a series of memory sectors within the same memory module. For example, this command is used to telemeter the memory sector in which the navigation update parameters are stored. The real-time TERMINATE command may by used to halt a sector dump before the last block of data is telemetered. The SECTOR DUMP command applies to the orbital phase.

The NAVIGATION UPDATE command permits loading of six navigation parameters and an execution time into the LVDC.

The SEQUENCE INITIATE UPDATE command permits update of stored values for the time of initiation of each of the preprogramed maneuvers specified for the mission. The prestored values will be adjusted immediately upon the receipt of the update command.

Other examples of command loads are: TIME BASE UPDATE, SLV PRELAUNCH TARGET UPDATE, and SLV ORBIT TARGET UPDATE.

DISPLAY AND CONTROL SYSTEM

MCC is equipped with facilities which provide for the input of data from the MSFN and KSC over a combination of high-speed data, low-speed data, wide-band data, teletype and television channels. This data is computer processed for display to the flight controllers. With this displayed data, detailed mission control by the MOCR and detailed support in the various specialty areas by the SSR's are made possible.

Display System

Several methods of displaying data are used including television (projection TV, group displays, closed circuit TV, and TV monitors), console digital readouts and event lights. The display and control system interfaces with the RTCC and includes computer request, encoder multiplexer, plotting display, slide file, digital-to-TV converter and telemetry event driver equipments (see figure 9-6).

The encoder multiplexer receives the display request from the console keyboard and encodes it into digital format for transmission to the RTCC.

The converter slide file data distributor routes slide selection data from the RTCC to reference slide files and converter slide files, receives RTCC control data signals required to generate individual console television displays and large scale projection displays, and distributes control signals to a video switching matrix to connect an input video channel with an output television viewer or projector channel.

The digital-to-television conversion is accomplished by processing the digital display data into alphanumeric, special symbol and vector displays for conversion into video signals. This process produces analog voltages which are applied to the appropriate element of a character-shaped beam cathode ray tube. The resultant display image on the face of the cathode ray tube is optically mixed with the slide file image and viewed by a television camera, which transmits the mixed images to the TV monitors and projectors. The digital-to-TV data consists of preprogramed, computer-generated, dynamic data formats which are processed and combined with background data on film slides.

Control System

A control system is provided for flight controllers to exercise their respective functions for mission control and technical management. This system is comprised of different groups of consoles with television monitors, request keyboards, communications equipment and assorted modules. These units are assembled as required, to provide each operational position in the MOCR with the control and display capabilities required for the particular mission. The console components are arranged to provide efficient operation and convenience for the flight controller. A typical console configuration for the Booster System Engineer is shown in figure 9-7. Brief descriptions of the console modules are contained in the following paragraphs.

The event indicator modules display discrete mission events, system modes, equipment modes, and vehicle system status. Each BSE event module consists of 18 bilevel indicators which permit a total of 36 event positions per module.

Another type of event module has the capability to display up to 72 events. The signals affecting the lights are telemetry inputs from the space vehicle.

The ground elapsed time module shows the elapsed time from liftoff.

The communications module provides rapid access to internal, external, and commercial voice communications circuits. Flight controllers may monitor as well as talk over these circuits.

The command module provides the flight controller with the means to select and initiate real-time commands and command loads for transmission to the space vehicle. The module also indicates receipt/rejection of commands to the space vehicle and verification of proper storage of command loads in a site command data processor. The modules are made up of pushbutton indicators (PBI's) which are labeled according to their function.

The status report module (SRM) provides flight controllers in the MOCR with the means to report mission status to the flight director and assistant flight director and to review systems status in the SSR's.

The abort request indicator provides the capability, by toggle switch action, to indicate an abort condition. This produces a priority command to the spacecraft.

The manual select keyboard (MSK) permits the flight controller to select a TV channel, a computer-generated display format, or a reference file for viewing on the TV monitor. The desired item is selected by use of a PBI three-mode switch (TV channel, reference file, display request) and a select-number thumbwheel encoder. BSE consoles may obtain a hardcopy of a display by use of a PBI. The hardcopy is delivered to the console via a pneumatic tube system.

The summary message enable keyboard (SMEK) permits the flight controller to instruct the RTCC to strip out selected data from telemetry inputs and to format this data into digital-to-TV summary displays. The SMEK is also used to instruct the RTCC to convert data into specific teletype messages to designated MSFN sites. The module contains appropriately labeled PBI's.

The TV monitor module provides viewing of computer-generated displays, reference file data, closed circuit TV within the MCC and KSC, and commercial TV.

Console modules not illustrated in figure 9-7 but used in MOCR and SSR consoles are described below.

The analog meter module displays parameters in engineering units. Up to 15 measurements, determined prior to the mission, can be displayed. There are movable markers on each meter which are manually set to show the nominal value

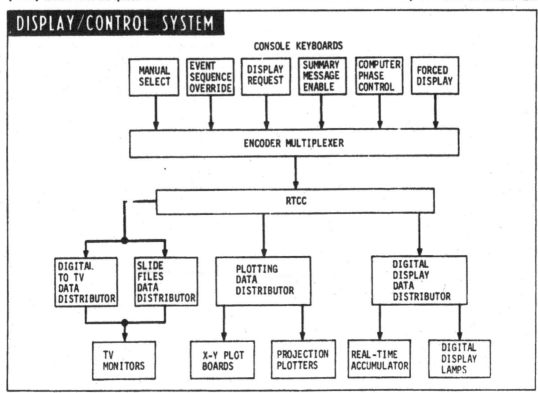

Figure 9-6

of the parameter as well as the upper and lower limits. When a parameter exceeds the established tolerance, a red warning light at the bottom of the meter is lit. The light extinguishes when the parameter returns to tolerance or the exceeded limit is manually extended.

The forced display module (FDK) indicates to the flight controller a violation of preprogramed limits of specific analog parameters as determined from the incoming data. When an out of tolerance condition occurs, the appropriate PBI in the module illuminates as a warning. When the flight controller acknowledges the lit PBI by depressing it, a readout indicator on the FDK will display a four digit code which identifies the display format on which the out of tolerance parameter appears.

The display request keyboard (DRK) provides a fast means of requesting the RTCC for a specific display format. The display is called up by depressing the appropriately labeled PBI. This keyboard provides the same capability as MSK in the "display request mode" except that the callup is faster in that thumbweel selection is not required.

CONTINGENCY PLANNING AND EXECUTION

Planning for a mission begins with the receipt of mission requirements and objectives. The planning activity results in specific plans for prelaunch and launch operations, preflight training and simulation, flight control procedures, flight crew activities, MSFN and MCC support, recovery operations, data acquisition and flow, and other mission related operations. Simulations are planned and performed to test procedures and train flight control and flight crew teams in normal and contingency operations. The simulation and training exercises result in a state of readiness for the mission. Mission documentation covering all aspects of the mission is developed and tested during the planning and training period. Included in this documentation are the mission rules.

MISSION RULES

Mission rules are a compilation of rules governing the treatment of contingency situations. The purpose of the mission rules is to outline preplanned actions to assist in making rapid real-time decisions during prelaunch, flight and recovery operations. The mission rules are based upon the mission objectives and on the objective of maintaining a high degree of confidence in crew safety during mission implementation. The mission rules categorize the degree of importance assigned to space vehicle/operational support elements as follows:

A mandatory item (M) is a space vehicle element or operational support element that is essential for accomplishment of the primary mission, which includes prelaunch, flight, and recovery operations that ensure crew safety and effective operational control as well as the attainment of the primary mission objectives.

A highly desirable item (HD) is a space vehicle or operational support element that supports and enhances the accomplishment of the primary mission or is essential for accomplishment of the secondary mission objectives.

Redline values are the maximum and/or minimum limits of a critical parameter (redline function) necessary to describe vehicle, system and component performance and operation. Redline functions are mandatory items.

Launch Mission Rules

Launch mission rules cover the following listed items and other information as appropriate.

1. Mandatory and highly desirable onboard instrumentation required to collect data for flight control purposes or postflight evaluation.

2. Mandatory and highly desirable onboard instrumentation required to verify that the space vehicle is ready for launch.

3. Redline values defining upper and lower limits of parameters, such as pressure, temperature, voltage, current and operating time, for any system/subsystem essential to mission success.

4. Mandatory and highly desirable range and instrumentation support required to prepare and launch the space vehicle and accomplish postflight analysis.

5. Range safety requirements and instrumentation as established by the Air Force Eastern Test Range.

6. Wind and weather restrictions on the launch.

7. Long-range camera coverage required for the launch.

8. Launch window definition and launch window rules pertaining to launch operations.

9. The space vehicle functional sequence.

10. The time span before launch during which manual cutoff will not be attempted.

Flight Mission Rules

Flight mission rules cover the following listed items and also medical decision rules for appropriate mission go/no-go points.

1. Mandatory and highly desirable instrumentation for control of the space vehicle after liftoff.

2. Space vehicle nominal and non-nominal subsystem performance in accordance with alternate mission capability.

3. Trajectory and guidance.

4. Mandatory and highly desirable items of mission support in the MCC.

5. Mandatory and highly desirable range and MSFN support required to support the mission after liftoff, and subsequent analysis and evaluation.

6. Rules relating to the human or medical aspects of manned flight.

7. Recovery restrictions.

8. Launch window rules pertaining to such items as time of liftoff, launch azimuth, recovery and spacecraft performance limitations.

VEHICLE MISSION ABORT CAPABILITY

Section III of this manual, dealing with the emergency detection system, describes the manual and automatic capabilities for mission abort designed into the Saturn-Apollo system. Also described in Section III are the abort modes and limits, and the emergency procedures related to mission abort.

Time critical aborts must be initiated onboard the spacecraft because sufficient time is not available for response by the MCC ground based flight controllers.

The detection of slowly diverging conditions which may result in an abort is the prime responsibility of MCC. In the event such conditions are discovered, MCC requests abort of the mission or, circumstances permitting, sends corrective commands to the vehicle or requests corrective flight crew action.

In the event of a non-catastrophic contingency, MCC recommends alternate flight procedures, and mission events are rescheduled to derive maximum benefit from the modified mission.

ABORT GROUND RULES

Flight crew safety shall take precedence over the accomplishment of mission objectives.

The Command Pilot of a manned mission may initiate such inflight action as he deems necessary for crew safety.

The Launch Operations Manager may send an abort request signal from the time the launch escape system (LES) is armed until the space vehicle reaches sufficient altitude to clear the top of the umbilical tower.

From liftoff to tower clear, the Launch Director and Flight Director have concurrent responsibility for sending an abort request.

Control of the space vehicle passes from the Launch Director to the Flight Director when the space vehicle clears the top of the tower.

In the Mission Control Center, the Flight Director, Flight Dynamics Officer, and Booster Systems Engineer have the capability to send an abort request signal.

Where possible, all manual abort requests from the ground during flight will be based on two independent indications of the failure.

LAUNCH VEHICLE MALFUNCTIONS/HAZARDS

Malfunctions which could result in loss of the space vehicle are analyzed and a mission rule is developed to respond to the malfunction. The malfunctions and corresponding hazards are incorporated into the premission simulations and the mission rules are refined as required. Typical malfunctions/hazards on which mission rules are based are listed in figure 9-8.

TYPICAL CONFIGURATION - BOOSTER SYSTEMS ENGINEERS CONSOLE

1. EVENT INDICATORS (S-IC/S-II)
2. EVENT INDICATOR (VEHICLE TELEMETRY STATUS, EDS, COMMAND, RANGE SAFETY)
3. EVENT INDICATORS (S-IVB)
4. GROUND ELAPSED TIME MODULE
5. EVENT INDICATORS (GUIDANCE AND NAVIGATION, ATTITUDE CONTROL)
6. EVENT INDICATORS (GUIDANCE AND NAVIGATION, ATTITUDE CONTROL)
7. COMMUNICATIONS MODULE
8. TV MONITOR
9. TV MONITOR
10. TV MONITOR
11. COMMUNICATIONS MODULE
12. COMMAND MODULE
13. COMMAND MODULE
14. COMMAND MODULE
15. STATUS REPORT MODULE
16. ABORT REQUEST MODULE
17. MANUAL SELECT KEYBOARD
18. COMMUNICATIONS MODULE
19. SUMMARY MESSAGE ENABLE KEYBOARD

Figure 9-7

TYPICAL LAUNCH VEHICLE MALFUNCTIONS/HAZARDS

Malfunction	Possible Hazard	Abort Requirement	Malfunction	Possible Hazard	Abort Requirement
	S-IC SUMMARY		Loss of Attitude Command	1. Failure in yaw and roll to large offset values causes loss of control 2. Failure in pitch to current value causes tilt arrest	Certain
Single Engine Out	1. Pad fallback up to 0.25 seconds 2. Holddown arm interference up to 0.9 seconds 3. Tower collision up to 5.6 seconds (tower side engine only)	Certain			
	4. Loss of control in max-q region with adverse winds	Possible	Saturated Rate Signal	1. Actuators hardover; loss of control	Certain
Actuator Hardover	1. Holddown arm interference up to 0.7 seconds 2. Tower Collision up to 3.2 seconds (plus yaw only)	Certain	Loss of Attitude Rate Signal	1. Rapid vehicle oscillation and buildup of excessive vehicle rates	Certain
	3. Loss of control in max-q region with 5-95 percentile winds	Possible	No S-IC/S-II First Plane Separation	1. No. S-II ignition possible	Certain
Dual Engine Out	1. Pad fallback up to 30 seconds 2. Holddown arm interference up to 1.5 seconds 3. Tower collision up to 7.3 seconds (2 tower side engines) 4. CSM joint capability exceeded up to 120 seconds (adjacent control engines) 5. Loss of control up to 60 seconds (2 opposite engines; or lower and center engines; or upper and center engines)	Certain	No S-II Second Plane Separation	1. Loss of vehicle due to over-heating	Certain
			Failure to Jettison LET	1. Requires use of emergency procedures for removal of LET after orbit is achieved	None
			No S-II/S-IVB Separation	1. Loss of mission; orbit is not possible	Certain
			PU System Failure	1. Primary effects are trajectory deviations and increased S-IVB first burn times	None
Saturated Control Signal	1. Pad fallback up to 30 seconds 2. Holddown arm interference up to 1.2 seconds 3. Tower collision up to 3.5 seconds for plus saturation in yaw and up to 6.5 seconds for negative yaw saturation 4. Loss of control (rapid tumbling) at all flight times and for all winds	Certain	One Actuator Hardover	1. For outboard case, no adverse effect	None
				2. For inboard case, may generate heating problem	Possible
			Single Engine Out	1. With an engine out at approximately 3.5 seconds prior to second plane separation, vehicle and interstage collision is possible	Possible
				2. For cases other than 1. above, deviation from nominal trajectory and increased S-IVB first burn times	None
Loss of Attitude Rate Signal	1. Oscillatory divergence of vehicle attitude at all flight times; loss of control	Certain			
Loss of Attitude Error Signal	1. Exponential divergence of vehicle attitude at all flight times; loss of control after 75 seconds	Certain			
	S-II SUMMARY		Dual Engine Out	1. Possible loss of control 2. Deviation from nominal trajectory and increased S-IVB first burn times	Possible
Loss of Attitude Error Signal	1. For pitch or yaw, loss of error signal causes slow divergence from nominal trajectory and, eventually, excessive attitude deviation	Probable		**S-IVB SUMMARY** (Applicable to both first and second burns except where noted)	
Saturated Error Signal	1. For pitch or yaw, causes actuators hardover in the affected plane; loss of control	Certain	Saturated Error Signal	1. For pitch or yaw, causes actuators hardover in affected plane; loss of control	Certain

Figure 9-8 (Sheet 1 of 2)

	TYPICAL LAUNCH VEHICLE MALFUNCTIONS/HAZARDS (Continued)				
Malfunction	Possible Hazard	Abort Requirement	Malfunction	Possible Hazard	Abort Requirement
Loss of Attitude Command	**S-IVB SUMMARY (Continued)** 1. Failure in yaw to large offset value causes loss of control	Certain	Loss of Both APS Modules	1. During powered flight (first burn), loss of roll control 2. During EPO, loss of attitude control	Certain
	2. Failure in pitch to current value causes tilt arrest	Certain (except for late portion of burn)		3. During powered flight (second burn), loss of roll control; if TLI is attained, vehicle rates will continue uncorrected, presenting difficulties for TD&E	Possible
Saturated Rate Signal	1. Actuators hardover; loss of control	Certain	Loss of One APS Module	1. During powered flight (first and second burns), causes attitude deviations and small changes in vehicle rates	None
Loss of Attitude Rate Signal	1. Rapid vehicle oscillation and buildup of excessive vehicle rates	Certain			
Loss of Attitude Error Signals	1. For pitch or yaw, causes slow divergence from nominal trajectory and, eventually, excessive attitude deviation	Probable		2. During EPO, loss of control in all three axes	Certain
				3. During TLI coast, vehicle rates may cause difficulties for TD&E	Possible
PU System Failure	1. Deviation from nominal trajectory.	None	Engine Out	1. During first burn, CSM insertion is possible if engine-out occurred after 10 minutes of flight time 2. During second burn, TLI capability determined by real-time analysis	Staging to SPS
One Actuator Hardover	1. Loss of control	Certain			
One Actuator Inoperative	1. Excessive attitude deviation if failure occurs prior to 30 seconds before nominal S-IVB cutoff	Certain			
			NOTE: Except where otherwise noted, time is measured from launch first motion		

Figure 9-8 (Sheet 2 of 2)

VEHICLE FLIGHT CONTROL PARAMETERS

In order to perform flight control monitoring functions, essential data must be collected, transmitted, processed, displayed, and evaluated to determine the space vehicle's capability to start or continue the mission. Representative parameters included in this essential data will be briefly described in the following paragraphs.

PARAMETERS MONITORED BY LCC

The launch vehicle checkout and prelaunch operations monitored by the Launch Control Center (LCC) were briefly discussed in Section VIII of this manual. These operations determine the state of readiness of the launch vehicle, ground support, telemetry, range safety, and other operational support systems. During the final countdown, hundreds of parameters are monitored to ascertain vehicle, system, and component performance capability. Among these parameters are the "redlines". The redline values must be within the predetermined limits or the countdown will be halted.

Typical redlines are fuel and oxidizer tank ullage pressure, GN_2 and helium storage sphere pressure, hydraulic supply pressures, thrust chamber jacket temperatures, bus voltages, IU guidance computer operations, H_2 and O_2 concentrations, and S-IVB oxidizer and fuel recirculation pump flow. In addition to the redlines, there are a number of operational support elements such as ALDS, range instrumentation, ground tracking and telemetry stations, ground communications, and other ground support facilities which must be operational at specified times in the countdown.

PARAMETERS MONITORED BY BOOSTER SYSTEMS GROUP

The Booster Systems Group monitors launch vehicle systems (S-IC, S-II, S-IVB, and IU) and advises the flight director and flight crew of any system anomalies. They are responsible for abort actions due to failure or loss of thrust and overrate conditions; for confirming inflight power, stage ignition, holddown release, all engines go, roll and pitch initiate, engine cutoffs, etc; for monitoring attitude control and stage separations; and for digital commanding of LV systems.

Specific responsibilities in the group are allocated as follows:

1. BSE No. 1 has overall responsibility for the group, for commands to the launch vehicle, and for monitoring and evaluating the S-IC and S-II flight performance. Typical flight control parameters monitored include engine combustion chamber pressure, engine gimbal system supply pressure, fuel and oxidizer tank ullage pressure, helium storage tank pressure, engine actuator (yaw/pitch/roll) position, THRUST OK pressure switches, longitudinal acceleration, vent valve positions, engine ignition/cutoff, and various bus voltages.

2. BSE No. 2 supports BSE No. 1 in monitoring the S-II flight and assumes responsibility for monitoring the S-IVB burns. Parameters monitored are similar to those monitored by BSE No. 1.

3. BSE No. 3 monitors the attitude control, electrical, guidance and navigation, and IU systems. Typical parameters monitored include roll/pitch/yaw guidance and gimbal angles; angular rates; ST-124 gimbal temperature and bearing pressure; LVDC temperature; and various bus voltages.

The preceding flight controllers are located in the MOCR. The following are located in the vehicle system SSR.

4. Guidance and Navigation Systems (GND) Engineer monitors the guidance, navigation, and digital (sequential) BSE systems. The GND provides detailed support to BSE No. 3. Typical parameters monitored includes ST-124-M3 accelerometer and gyro pickups (X,Y,Z axes); and fixed position and fixed velocity (X,Y,Z components).

5. Attitude Control and Stabilization Systems (ACS) Engineer monitors the attitude control system, the S-IVB hydraulic and auxiliary propulsion systems, and the emergency detection system. The ACS provides detailed support to BSE's No. 2 and 3. Typical parameters monitored include hydraulic accumulator pressures, hydraulic reservoir piston position, attitude control fuel and oxidizer module temperatures, and excessive rates (pitch/roll/yaw).

6. Engine Systems Engineer monitors the S-II and S-IVB engine systems and O_2/H_2 burner and provides detailed support to BSE No. 2. Among the parameters monitored are thrust chamber pressure, engine inlet lox and LH_2 pressure, pre-valve position, and O_2/H_2 burner chamber dome temperature.

7. Stage Systems Engineer monitors the pressurization, repressurization, bulkhead pressure differential and chilldown and provides detailed support to the BSE No. 2. Typical parameters monitored include helium tank pressure, start tank pressure/temperature, and common bulkhead pressure.

8. Electrical Network and Systems (ENS) Engineer monitors electrical systems (all LV stages), IU environmental control system, and range safety systems (safing at orbital insertion). The ENS

provides detailed support to BSE No. 3. Typical parameters monitored include exploding bridgewire voltages, sublimator inlet temperature, GN_2 regulator inlet temperature, and various bus voltages and currents.

9. Command Systems Engineer monitors the commands sent to the launch vehicle and advises BSE No. 1 on their status. In the event of rejection of a command by the onboard computer, he determines the cause of the rejection, i.e., improperly coded command, malfunction of the command system, malfunction of the computer, etc.

10. Consumables Engineer monitors status at all times of launch vehicle consumables including all high pressure spheres, APS propellants, and main stage propellants. He advises the BSE No. 2 of mission impact when consumables are depleted beyond predicted limits.

PARAMETERS MONITORED BY FLIGHT DYNAMICS GROUP

The Flight Dynamics Group monitors and evaluates the powered flight trajectory and makes the abort decisions based on trajectory violations. It is responsible for abort planning, entry time and orbital maneuver determinations, rendezvous planning, inertial alignment correlation, landing point prediction, and digital commanding of the guidance systems.

The MOCR positions of the Flight Dynamics Group include the Flight Dynamics Officer (FDO), the Guidance Officer (GUIDO), and the Retrofire Officer (RETRO). The MOCR positions are given detailed specialized support by the flight dynamics SSR.

The surveillance parameters measured by the ground tracking stations and transmitted to the MCC are computer processed into plotboard and digital displays. The flight dynamics group compares the actual data with premission calculated nominal data and is able to determine mission status. The surveillance parameters include slant range, azimuth and elevation angles, antenna polarization angle, and other data. From these measurements, space vehicle position, velocity, flight path angle, trajectory, ephemeris, etc., may be calculated. Typical plotboard displays generated from the surveillance parameters are altitude versus downrange distance, latitude versus longitude, flight path angle versus inertial velocity, and latitude versus flight path angle.

PARAMETERS MONITORED BY SPACECRAFT SYSTEMS GROUP

The Spacecraft Systems Group monitors and evaluates the performance of spacecraft electrical, optical, mechanical, and life support systems; maintains and analyzes consumables status; prepares mission log; coordinates telemetry playback; determines spacecraft weight and center of gravity; and executes digital commanding of spacecraft systems.

The MOCR positions of this group include the Command and Service Module Electrical, Environmental, and Communications Engineer (CSM EECOM), the CSM Guidance, Navigation, and Control Engineer (CSM GNC), the Lunar Module Electrical, Environmental, and

Communications Engineer (LM EECOM), and the LM Guidance, Navigation, and Control Engineer (LM GNC). These positions are backed up with detailed support from the vehicle systems SSR.

Typical parameters monitored by this group include fuel cell skin and condenser temperatures, fuel cell current, various battery and bus voltages, launch escape tower and motor discretes, AGCU drift, SPS helium tank pressure, SPS fuel and oxidizer tank pressure, and fuel and oxidizer inlet pressure differential.

PARAMETERS MONITORED BY LIFE SYSTEMS GROUP

The Life Systems Group is responsible for the well being of the flight crew. The group is headed by the Flight Surgeon in the MOCR. Aeromedical and environmental control specialists in the life systems SSR provide detailed support to the Flight Surgeon. The group monitors the flight crew health status and environmental/biomedical parameters.

MANNED SPACE FLIGHT NETWORK

The Manned Space Flight Network (MSFN) is a global network of ground stations, ships, and aircraft designed to support manned and unmanned space flight. The network provides tracking, telemetry, voice and teletype communications, command, recording, and television capabilities. The network is specifically configured to meet the requirements of each mission.

MSFN stations are categorized as lunar support stations (deep-space tracking in excess of 15,000 miles), near-space support stations with Unified S-Band (USB) equipment, and near-space support stations without USB equipment. Figure 9-9 is a matrix listing the stations, designators, and capabilities. Figure 9-10 shows the geographical location of each station.

MSFN stations include facilities operated by NASA, the United States Department of Defense (DOD), and the Australian Department of Supply (DOS).

The DOD facilities include the Eastern Test Range (ETR), Western Test Range (WTR), White Sands Missile Range (WMSR), Range Instrumentation Ships (RIS), and Apollo Range Instrumentation Aircraft (ARIA). Recovery forces under DOD are not considered to be part of the MSFN.

NASA COMMUNICATIONS NETWORKS

The NASA Communications (NASCOM) network is a point-to-point communications systems connecting the MSFN stations to the MCC. NASCOM is managed by the Goddard Space Flight Center, where the primary communications switching center is located. Three smaller NASCOM switching centers are located at London, Honolulu, and Canberra. Patrick AFB, Florida and Wheeler AFB, Hawaii serve as switching centers for the DOD eastern and western test ranges, respectively. The MSFN stations throughout the world are interconnected by landline, undersea cable, radio and communications satellite circuits. These circuits carry teletype, voice, and data in real-time support of the missions. Figure 9-11 depicts a typical NASCOM configuration.

Each MSFN USB land station has a minimum of five

voice/data circuits and two teletype circuits. The Apollo insertion and injection ships have a similar capability through the communications satellites.

The Apollo Launch Data System (ALDS) between KSC and MSC is controlled by MSC and is not routed through GSFC. The ALDS consists of wide-band telemetry, voice coordination circuits, and a high speed circuit for the Countdown and Status Transmission System (CASTS). In addition, other circuits are provided for launch coordination, tracking data, simulations, public information, television, and recovery.

MSFC SUPPORT OF LAUNCH AND FLIGHT OPERATIONS

The Marshall Space Flight Center (MSFC) by means of the Launch Information Exchange Facility (LIEF) and the Huntsville Operations Support (HOSC) provides real-time support of launch vehicle prelaunch, launch, and flight operations. MSFC also provides support via LIEF for postflight data delivery and evaluation.

LAUNCH INFORMATION EXCHANGE FACILITY

The LIEF encompasses those personnel, communications, data processing, display, and related facilities used by the MSFC launch vehicle design team to support Apollo-Saturn mission operations. The LIEF operations support organization is shown in figure 9-12.

In-depth real-time support is provided for prelaunch, launch, and flight operations from HOSC consoles manned by engineers who perform detailed system data monitoring and analysis.

System support engineers from MSFC and stage contractors are organized into preselected subsystem problem groups (approximately 160 engineers in 55 groups) to support KSC and MSC in launch vehicle areas which may be the subject of a request for analysis. The capabilities of MSFC laboratories and the System Development Facility (SDF) are also available.

PRELAUNCH WIND MONITORING

Prelaunch flight wind monitoring analyses and trajectory simulations are jointly performed by MSFC and MSC personnel located at MSFC during the terminal countdown. Beginning at T-24 hours, actual wind data is transmitted periodically from KSC to the HOSC. These data are used by the MSFC/MSC wind monitoring team in vehicle flight digital simulations to verify the capability of the vehicle with these winds. Angle of attack, engine deflections and structural loads are calculated and compared against vehicle limits. Simulations are made on either the IBM 7094 or B5500 computer and results are reported to the Launch Control Center (LCC) within 60 minutes after wind data transmission. At T-2 1/4 hours, a go/no-go recommendation is transmitted to KSC by the wind monitoring team. A go/no-go condition is also relayed to the LCC for lox and LH_2 loading.

In the event of marginal wind conditions, contingency wind data balloon releases are made by KSC on an hourly basis after T-2 1/2 hours and a go/no-go recommendation is transmitted to KSC for each contingency release. This contingency data is provided MSFC in real-time via CIF/DATA-CORE and trajectory simulations are performed on-line to expedite reporting to KSC.

Ground wind monitoring activities are also performed by MSFC laboratory personnel for developmental tests of displays. Wind anemometer and strain gauge data are received in real-time and bending moments are computed and compared with similar bending moment displays in CIF.

LAUNCH AND FLIGHT OPERATIONS SUPPORT

During the prelaunch period, primary support is directed to KSC. Voice communications are also maintained with the Booster Systems Group at MCC and the KSC/MCC support engineers in CIF to coordinate preparations for the flight phase and answer any support request.

At liftoff, primary support transfers from KSC to the MCC. The HOSC engineering consoles provide support, as required, to the Booster Systems Group for S-IVB/IU orbital operations, by monitoring detailed instrumentation for system inflight and dynamic trends, by assisting in the detection and isolation of vehicle malfunctions, and for providing advisory contact with vehicle design specialists. This support is normally provided from liftoff through the active launch vehicle post-spacecraft separation phase, approximately T + 6 hours, or until LIEF mission support termination.

MSFN STATIONS/EQUIPMENT MATRIX

Facility	Designation	85' Dual	30' Single	30' Dual	12' Single	7' USB	7' Dish	AGAVE	TELTRAC	TLM-18 (60')	30' Dish	12' X-Dipole	C-Band Radar	High Speed Tracking Data	A/G Remote (VHF)	RSO Displays	High Speed TM Data	IU Command	IU Telemetry	Voice SRD	Voice SCAMA	Teletype	Apollo RSDP'S	Apollo TV Monitor	TV Scan Converter
Mission Control Center, Houston	MCC																			X	X	X			
MSFN Operations Center, Goddard	MSFNOC																			X	X	X			
Air Operations Control Center, PAFB	AOCC																			X	X	X			
Central Instrumentation Facility/ Apollo Launch Data System, KSC	CIF/ALDS													X											
Ascension Island (NASA) *	ACN				X				X					X	X		X	X	X	X	X	X	X	X	
Antigua (NASA) *	ANG		X						X					X	X		X			X	X	X	X	X	
Antigua (DOD-ETR) *	ANT									X	X	X	X	X		X				X	X	X			
Ascension Island (DOD-ETR) *	ASC									X	X	X	X	X		X				X	X	X			
Bermuda (NASA) *	BDA		X						X	X	X			X	X	X	X	X	X	X	X	X	X	X	
Vandenberg, California (DOD-WTR) *	CAL													X		X				X	X	X			
Honeysuckle Creek, Australia (WRE) **	HSK	X												X		X	X		X	X	X	X	X	X	X
Honeysuckle Creek, Australia (6) **	HSKX	X												X											
Cape Kennedy (DOD-ETR) *	CNV													X		X		X	X	X		X			
Carnarvon, Australia (WRE) *	CRO			X										X	X	X	X	X	X	X	X	X	X	X	
Canary Islands (NASA) *	CYI		X						X	X				X	X	X	X			X	X	X	X	X	
Grand Bahama Island (DOD-ETR) *	GBI									X	X	X	X	X		X				X		X			
Grand Turk Island (DOD-ETR) ***	GTK												X	X						X	X	X			
Grand Bahamas (NASA) *	GBM		X						X					X	X	X	X			X		X	X	X	
Goldstone, California (NASA) **	GDS	X												X		X	X		X	X	X	X	X	X	X
Goldstone, California (6) **	GDSK	X												X											
Guam (NASA) *	GWM		X											X	X	X	X			X	X	X	X	X	
Guaymas, Mexico (NASA) *	GYM		X						X					X	X	X	X			X	X	X	X	X	
Kokee Park, Hawaii (NASA) *	HAW			X		X	X							X	X	X	X			X	X	X	X	X	
Madrid, Spain (NASA) **	MAD	X												X		X	X		X	X	X	X	X	X	X
Madrid, Spain (6) **	MADX	X												X											
Merrit Island Florida (NASA) *	MIL		X						X					X	X	X	X	X		X	X	X	X	X	
Merritt Island, Florida (DOD-ETR) *	MLA									X	X			X						X		X			
Patrick AFB, Florida (DOD-ETR) *	PAT									X	X			X						X		X			
Pretoria, So. Africa (DOD-ETR) *	PRE											X	X	X						X		X			
Tananarive, Malagasy (NASA) *	TAN								X					X						X		X			
Corpus Christi, Texas (NASA) *	TEX		X						X	X				X	X	X	X	X		X	X	X	X	X	
Apollo Range Instrumented Aircraft *	ARIA					X	X							X						X	X	X	X	X	
USNS Huntsville (DOD) *	HTV (Ship)				X								X	X						X	X	X			
USNS Mercury (DOD) *	MER (Ship)		X										X	X	X	X	X			X	X	X	X	X	
USNS Redstone (DOD) *	RED (Ship)		X										X	X	X	X	X			X	X	X	X	X	
USNS Vanguard (DOD) *	VAN (Ship)		X										X	X	X	X	X			X	X	X	X	X	

*   Near Space Support Stations

**   Lunar Support Stations

*** Supplemental Stations (not listed above are Space Defense Center (NORAD) and NASA facilities which are used as required)

① Unified S-Band antenna for telemetry, voice, range and range rate, updata, and tracking.

② USNS Huntsville does not have updata capability

③ ARIA do not have update or range and range rate capability

④ VHF antenna for telemetry and tracking

⑤ Apollo telemetry and command remote site data processors (UNIVAC 542B)

⑥ MSFN wings of Jet Propulsion Laboratory buildings at these Lunar Support Stations

Figure 9-9

GEOGRAPHIC LOCATIONS, MSFN STATIONS

Figure 9-10

Figure 9-11

LIEF OPERATIONS SUPPORT ORGANIZATION

Figure 9-12

This Space Intentionally Left Blank

MISSION VARIABLES AND CONSTRAINTS

TABLE OF CONTENTS

INTRODUCTION

The Apollo 12, AS-507, H-1 mission is an Apollo Lunar Surface Experiments Package (ALSEP) manned Lunar Landing Mission (LLM) to be implemented during the November, 1969 launch month (11-16 November). The primary launch day is November 14 and the backup launch day is November 16.

Lunar landing missions of the H-type are planned for Apollo 13/SA-508, Apollo 14/SA-509 and Apollo 15/SA-510 space vehicles.

MISSION OBJECTIVES

The H-1 mission primary objectives are as follows:

1. Perform selenological inspection, survey, and sampling in a mare area.

2. Deploy an ALSEP consistent with a seismic net.

3. Develop techniques for a point landing.

4. Develop man's capability to work in the lunar environment.

The LV major flight objectives are:

1. Demonstrate the LV capability to inject the spacecraft (SC) into the specified translunar trajectory.

2. Demonstrate LV capability to maintain a specified attitude for transportation, docking, and lunar module ejection (TD&E) operations after translunar injection (TLI).

3. Demonstrate LV capability to perform the S-IVB evasive maneuver.

4. Perform LV dumping and safing after final LV/SC separation and the S-IVB evasive maneuver to reduce the probability of LV (S-IVB, instrument unit, and spacecraft/lunar module adapter) recontact with the SC, LV earth impact, or LV lunar impact, in that order of priority.

The SC major flight objectives are:

1. Demonstrate acceptable service propulsion system (SPS) performance and SC guidance during the lunar orbit insertion boost and the transearth injection boost.

2. Demonstrate acceptable LM systems performance during the descent-to-hover boost.

3. Demonstrate acceptable LM systems performance during the ascent and rendezvous mode.

Mission objectives for the H-type missions are similar to the objectives of the H-1 mission.

MISSION PROFILE

The mission profile for the H-1 mission consists of the following phases:

1. Launch to earth parking orbit (EPO).

2. Translunar injection (TLI).

3. Translunar coast.

4. S-IVB evasive maneuver.

5. S-IVB slingshot.

6. Lunar orbit insertion.

7. Lunar module (LM) descent.

8. Lunar surface operations.

9. LM ascent.

10. Transearth injection.

11. Transearth coast.

12. Entry and recovery.

The mission profile for a lunar landing mission is shown on figure 10-1. Figure 10-2 gives the operational trajectory ground track for the AS-507/H-1 mission for the November 14, 1969 launch date.

Descriptions of the mission phases are contained in the following paragraphs.

LAUNCH TO EPO

The launch vehicle places the spacecraft with three crewmen aboard into a 100 nautical mile circular parking orbit. The boost to EPO consist of complete burns of the S-IC and S-II stages and a partial burn of the S-IVB stage.

The vehicle coasts in EPO for up to three revolutions while launch vehicle and spacecraft subsystems checkout is performed. During coast in EPO, no major thrusting occurs; however, the orbit is continuously perturbed by low-level LH_2 venting thrust. Auxiliary propulsion system (APS) ullage burns, aerodynamic drag and earth oblateness further perturb the orbit. Launch vehicle attitude, vent, and propellant dump timelines for EPO insertion, TLI, and translunar coast, are shown in figure 10-3.

OPERATIONAL TRAJECTORY GROUNDTRACK

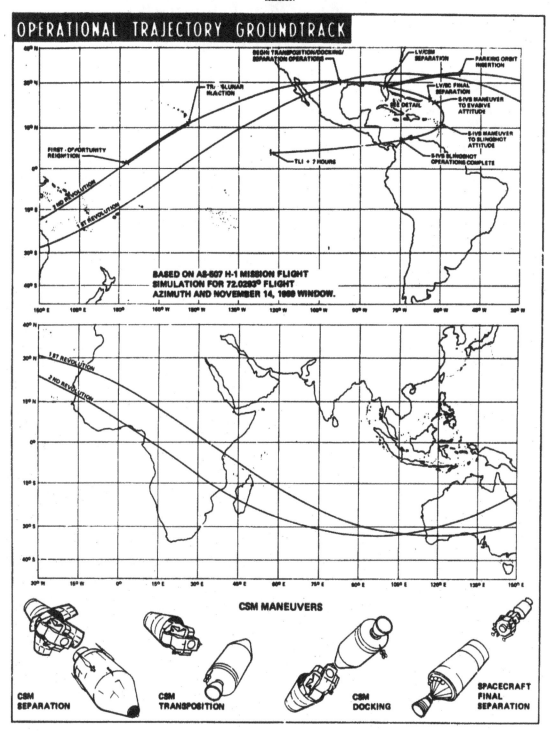

BASED ON AS-507 H-1 MISSION FLIGHT
SIMULATION FOR 72.0283° FLIGHT
AZIMUTH AND NOVEMBER 14, 1969 WINDOW.

CSM MANEUVERS

CSM
SEPARATION

CSM
TRANSPOSITION

CSM
DOCKING

SPACECRAFT
FINAL
SEPARATION

Figure 10-2

ATTITUDE, VENT, AND PROPELLANT DUMP TIMELINES

EARTH PARKING ORBIT-TRANSLUNAR INJECTION

POSITION 1

S-IVB CUTOFF (FIRST BURN)	ALIGN ALONG LOCAL HORIZONTAL	S-IVB RESTART IGNITION	S-IVB CUTOFF (SECOND BURN)	TLI	ALIGN ALONG LOCAL HORIZONTAL
T_5-0:00:00.2	T_5+0:00:20.0	T_6+0:09:38.0	T_7-0:00:00.2	T_7+0:00:09.8	T_7+0:00:20.0

APS T_5+0:00:00.3 — T_5+0:01:27.0

LH₂ CV T_5+0:00:59.0 — T_6+0:00:42.2 ; T_7+0:00:00.5 — T_7+0:14:59.8

LH₂ NPV T_7+0:00:00.8 — T_7+0:14:59.0

LOX NPV T_7+0:00:00.7 — T_7+0:02:30.7

TRANSLUNAR TRAJECTORY

MANEUVER TO TD&E ATTITUDE	CSM SEPARATION	CSM DOCK	SC/LV SEPARATION	MANEUVER TO S-IVB EVASIVE ATTITUDE	START S-IVB EVASIVE BURN	START SLINGSHOT SEQUENCE
T_7+0:15:00.0	T_7+0:25:00.0	T_7+0:35:00.0	T_7+1:25:00.0	T_7+1:28:00.0	T_8+0:00:01.2	T_8+0:09:40.0

APS T_8+0:00:01.2 — T_8+0:01:21.2

LH₂ CV

LH₂ NPV T_7+1:00:00.4 — T_7+1:14:59.0

LOX NPV

S-IVB SLINGSHOT SEQUENCE

LOX DUMP T_8+0:21:20.0 — T_8+0:22:18.0

APS T_8+0:56:40.0 — T_8+1:01:40.0

LH₂ CV T_8+0:09:40.2 —

LH₂ NPV T_8+0:23:25.0 —

LOX NPV T_8+0:23:20.2 —

VENTS REMAIN OPEN

LEGEND:
- — — LOCAL HORIZONTAL
- ——— VENT CLOSED/NO DUMP/ULLAGE ENGINES OFF
- ▬▬▬ VENT OPEN/DUMP ON/ULLAGE ENGINES ON
- CV-CONTINUOUS VENT (PROPULSIVE)
- NPV-NON-PROPULSIVE VENT
- APS-S-IVB APS ULLAGE ENGINES

ATTITUDE	①	②	③
PITCH	+120°	+120°	+191°
ROLL	+180°	+180°	+180°
YAW	-40°	+40°	0°

TIME IN HOURS: MINUTES: SECONDS
FOR NOMINAL GROUND ELAPSED TIMES
SEE FIGURE 2-1, SECTION II

Figure 10 - 3

TRANSLUNAR INJECTION

The S-IVB stage is reignited to inject the space vehicle to the translunar conic. The reignition time and orbital position of the S-IVB injection boost depend upon the lunar declination and upon the injection window and opportunity to be used. The first injection opportunity occurs midway through the second revolution in parking orbit and the second injection opportunity occurs midway through the third revolution.

Ground traces for TLI boosts are shown on figure 10-4.

TRANSLUNAR COAST

Following S-IVB cutoff, the translunar orbit is perturbed by a propulsive hydrogen vent for approximately 15 minutes. After the propulsive vent closes, the vehicle maneuvers to the commanded TD&E attitude, and this attitude is maintained intertially throughout TD&E operations. The Command and Service Module (CSM) is separated from the reminder of the vehicle (S-IVB, IU, SC/LM adapter (SLA), and LM) by use of the Service Module (SM) reaction control system (RCS). The CSM transposes and docks with the LM. Separation of the combined LM/CSM from the S-IVB/IU/SLA is achieved by use of the SM RCS.

Translunar trajectories are classified as nominal or hybrid. The nominal trajectory places the SC into a low-periselenum free-return trajectory. This returns the SC to earth with no additional thrusting should lunar orbit insertion not be executed.

In the hybrid trajectory, the targeting objective at TLI is for a high-periselenum, free-return trajectory. High-periselenum altitudes are those greater than 100 nautical miles. Shortly after SC/LV final separation, the SC maneuvers to a non-free-return trajectory. This trajectory permits, should lunar orbit insertion not be executed, a maneuver to place the SC into a proper transearth trajectory.

Targeting objectives for the November 14 and 16 launch days are approximately 1,850 and 1,000 nautical miles, respectively.

S-IVB EVASIVE MANEUVER

Following the separation of the combined CSM/LM from the S-IVB/IU/SLA, the S-IVB/IU/SLA is maneuvered from the TD&E attitude to the S-IVB evasive maneuver attitude. In this maneuver, the vehicle yaw attitude angle is changed in sign but not in magnitude, for example, the yaw angle changes from -40 to +40 degrees; the pitch and roll angles remain the same. (See figure 10-3.) An 80-second burn of the S-IVB APS ullage engines propels the S-IVB/IU/SLA a safe distance away from the spacecraft.

S-IVB SLINGSHOT

After the S-IVB evasive maneuver is completed, the S-IVB slingshot procedure is initiated by maneuvering the the LV (S-IVB/IU/SLA) to the slingshot attitude. (See figure 10-3). A retrograde velocity differential is attained by a liquid oxygen dump through the J-2 engine, hydrogen propulsive venting, and a burn of the S-IVB APS ullage engines. This reduces vehicle velocity and changes the trajectory so that the LV coasts past the trailing side of the moon. The retrograde velocity differential is a variable depending on

several factors including launch time and date, flight azimuth, and TLI opportunity. On the trailing side, the LV is co-rotational with the moon. The moon's gravitational field increases LV velocity sufficiently to place it into solar orbit. (See figure 10-1). Following the retrograde dump of propellants, the S-IVB stage is safed by venting the remaining propellant and high pressure gas bottles.

LUNAR ORBIT INSERTION

The SM propulsion system is used to deboost the spacecraft into lunar orbit. A trim burn for plane change and/or circularization may be made if required. The lunar orbit altitude is approximately 60 nautical miles.

LM DESCENT

Two astronauts enter the LM, perform LM checkout, and separate from the CSM using the SM RCS. The LM descent propulsion system (DPS) is used for the descent to the lunar surface.

The lunar surface at the landing site is in the astronauts' field of view for at least 120 seconds prior to touchdown. At the time of landing, the sun elevation referenced to local horizontal at the landing site is between 5 and 13 degrees.

Lunar landing sites for the H-1 mission are shown in the following tabulation.

LAUNCH DATE	LUNAR LATITUDE	LUNAR LONGITUDE
November 14	2° 24' S	23° 18' W
November 16	1° 41' N	41° 54' W

LUNAR SURFACE OPERATIONS

The staytime on the lunar surface is open-ended up to a planned maximum of 32 hours. The nominal plan provides for two exploration periods of open-ended duration up to 3 hours each for simultaneous exploration by both astronauts with a maximum radius of operation limited to 1500 feet. After checkout of the LM to assess its launch capability, the LM may be depressurized to allow egress to the surface. The planned lunar surface activities include the following major items in tentative order of priority:

1. Photography through the LM window.

2. Collection of a contingency lunar soil sample.

3. Assessment of astronaut capabilities and limitations.

4. LM inspection.

5. Experiment deployment (ALSEP).

6. Lunar field geology, including collection of a documented lunar soil sample.

Photographic records will be obtained and extravehicular activity will be televised.

LUNAR MODULE ASCENT

Prior to LM liftoff, the CSM makes the required plane change

GROUND TRACES-TLI BOOSTS

LOCI OF INJECTION POSITIONS

108°   108°   90°   90°   72°   72°

LOCI OF IGNITION POSITIONS

——— FIRST OPPORTUNITY ENVELOPE

—·—·— SECOND OPPORTUNITY ENVELOPE

BASED ON AS-507 H-1 MISSION FLIGHT SIMULATIONS FOR NOVEMBER 14, 1969 LAUNCH DATE.

Figure 10-4

to permit a nominally coplanar rendezvous. The LM ascent propulsion system is used to achieve the initial orbit. The LM RCS is used to make a plane change trim burn (if required), to achieve a concentric orbit, to transfer to the CSM orbit, and to rendezvous and dock with the CSM. Once docked to the CSM, the two LM crewmen transfer to the CSM. The CSM is separated from the LM using the SM RCS.

TRANSEARTH INJECTION

The SM propulsion system is used to boost the CSM out of lunar orbit. The nominal flight duration does not exceed 110 hours, and the return inclination does not exceed 40 degrees relative to the earth's equator.

TRANSEARTH COAST

During transearth coast, midcourse corrections are made, if required, using the Manned Space Flight Network for navigation.

ENTRY AND RECOVERY

Prior to atmospheric entry, the CM separates from the SM using the SM RCS. CM control during the atmospheric portion of flight is maintained by the roll position that controls the lift-to-drag ratio. Control in this manner provides a maneuvering capability through the reentry corridor.

Earth touchdown is in the Pacific within ±35 degrees latitude

and nominally occurs within 11 days of the launch from earth.

Recovery and transport of the crew and equipment to the Lunar Receiving Laboratory will be conducted under approved quarantine procedures.

MISSION CONSTRAINTS

Mission requirements impose the following constraints on the launch vehicle, spacecraft, operations and planning:

1. Nominal acceleration during the S-IC boost shall not exceed 4.0 g's.

2. The propellant utilization systems for the S-II and S-IVB stages will be open loop.

3. The Launch Vehicle Digital Computer (LVDC) flight program will have the capability to conduct alternate missions. Contingencies and corresponding alternate missions will be identified in the Flight Mission Rules.

4. The LVDC flight program must be capable of accepting real-time Digital Computer System (DCS) commands for specific functions (except when inhibited by the crew) during earth orbital and TLI coast periods. Representative of these functions are navigation and target update, S-IVB or IU

generalized switch selector events, termination of DCS routines, maneuver to evasive attitude, and Time Base 8 enable. Utilization of these DCS mode commands will be as identified in the Flight Mission Rules.

5. In earth orbit and translunar coast modes, the crew has a manual attitude control capability. Crew commanded pitch and yaw attitude rates will be limited to 0.3 degrees per second, and roll rates to 0.5 degrees per second. Attitude control will revert to the launch vehicle when relinquished by the crew.

6. The spacecraft will provide attitude commands to the LVDC in the event of launch vehicle attitude reference failure during either powered flight or coast periods. Attitude commands must be provided by the spacecraft from the time of takeover until SC/LV separation. Utilization of this capability will be as identified in the Flight Mission Rules.

7. The LVDC flight program shall be capable of executing the following commands from the spacecraft: early S-II/S-IVB staging, inhibit S-IVB

TLI burn, and initiate S-IVB restart sequence (Time Base 6). Utilization of this capability will be as identified in the Flight Mission Rules.

8. Continuous S-IVB tracking, telemetry, and voice are required for a 10-minute period within the first 20 minutes after TLI cutoff.

9. Telemetry is required from the S-IVB for a minimum of 10 minutes following TLI. Tracking is required on a basis of non-interference with CSM support.

10. S-IVB tracking, telemetry, and command are required during the TD&E maneuver.

11. Rate limit in pitch, yaw and roll is one degree per second during the TD&E maneuver.

12. The flight azimuth shall be not less than 72 degrees and not greater than 108 degrees.

13. TLI will be performed during the second revolution of EPO. If system status precludes nominal injection during the second revolution, injection during the third revolution of EPO will be possible.

This Space Intentionally Left Blank

APPENDIX A
ABBREVIATIONS AND ACRONYMS

A

A .... Astronaut
ac .... Alternating Current
ACCEL .... Acceleration
ACE .... Acceptance Checkout Equipment
ACM .... Actuation Control Module
ACN .... NASA MSFN Station, Ascension Island
ACS .... Attitude Control and Stabilization Systems
                                Engineer (Booster Systems)
AF .... Air Force
AFB .... Air Force Base
AFD .... Assistant Flight Director
AFETR .... Air Force Eastern Test Range
AGAVE .... Automatic Gimbaled Antenna
                                Vectoring Equipment
AGC .... Apollo Guidance Computer
A/G COMM .... Air to Ground and Ground to Air
                                Communications
AGCU .... Apollo Guidance Control Unit
ALDS .... Apollo Launch Data System
ALSEP .... Apollo Lunar Surface Experiments Package
Alt .... Altitude
AM .... Amplitude Modulation
amp .... Ampere or Amplifier
AMR .... Atlantic Missile Range
ANG .... NASA MSFN Station, Antigua
Ant .... Antenna
ANT .... DOD-ETR MSFN Station, Antigua
AOA .... Angle of Attack
AOCC .... Air Operations Control Center, PAFB
AOS .... Acquisition of Signal
AOS .... Atlantic Ocean Ships
APS .... Auxiliary Propulsion System
ARIA .... Apollo Range Instrumentation Aircraft
ASC .... DOD-ETR MSFN Station, Ascension Island
ASD .... Abort Summary Document
ASI .... Augmented Spark Igniter
ATT .... Attitude
AUTO .... Automatic

B

BDA .... NASA MSFN Station, Bermuda
BIOMED .... Biomedical
BMAG .... Body Mounted Attitude Gyro
BEF .... Blunt End Forward
BSE .... Booster Systems Engineer

C

CAL DOD-WTR MSFN Station, Vandenberg AFB, California
calips .... Calibrationable Pressure Switch
CAPCOM .... Spacecraft Communicator
CASTS .... Countdown and Status Transmission System
CCATS . Command, Communications and Telemetry System
CCF .... Converter/Compressor Facility
CCS .... Command Communications System
CCW .... Counterclockwise
CDDT .... Countdown Demonstration Test
CDF .... Confined Detonating Fuse

CDP .... Command Data Processor, MSFN Site
CDR .... Critical Design Review
CDR .... Spacecraft Commander
CECO .... Center Engine Cutoff
CG .... Center of Gravity
CIF .... Central Instrumentation Facility
                                (Located at Kennedy Space Center)
CIU .... Computer Interface Unit
CKAFS .... Cape Kennedy Air Force Stations
CM .... Command Module
CMC .... Command Module Computer
CMD .... Command
CNTR .... Center
CNV .... DOD-ETR MSFN Station, Cape Kennedy
CO .... Cutoff or Checkout
COFW .... Certificate of Flight Worthiness
COI .... Contingency Orbit Insertion
COM .... Common
cps .... Cycles per Second (Hertz)
CRO .... WRE MSFN Station, Carnarvon
CRT .... Cathode Ray Tube
CSM .... Command and Service Module
C-T .... Crawler-Transporter
C&W .... Cable and Wireless
CW .... Clockwise
CW .... Continuous Wave
CYI .... NASA MSFN Station, Canary Islands
CZR .... High Speed Metric Camera

D

DATA-CORE .... CIF Telemetry Conversion System
db .... Decibel
dbm .... Decibels Referenced to One Milliwatt
dc .... Direct Current
DCR .... Destruct Command Receiver
DCR .... Design Certification Review
DCS .... Digital Computer System
DDAS .... Digital Data Acquisition System
DEE .... Digital Event Evaluator
deg .... Degree
DOD .... Department of Defense
DOS .... Department of Supply (Australia)
DPS .... Descent Propulsion System
DRK .... Display Request Keyboard
DSKY .... Display and Keyboard (Spacecraft Guidance
                                and Control)
DTS .... Data Transmission System

E

EAO .... Experiment Activities Officer
EBW .... Exploding Bridgewire
ECA .... Electrical Control Assembly
ECS .... Environmental Control System
ECU .... Environmental Control Unit
EDS .... Emergency Detection System
EECOM ... Electrical, Environmental, and Communications
                                Engineer
ELS .... Earth Landing System
EMS .... Entry Monitor System

ENG .... Engine
ENS .... Electrical Network and Systems Engineer
(Booster Systems)
EPO .... Earth Parking Orbit
ESE .... Electrical Support Equipment
ETR .... Eastern Test Range
EVA .... Extravehicular Activity
EVT .... Extravehicular Transfer

F

F .... Fahrenheit
F .... Force
FACI .... First Article Configuration Inspection
FAO .... Flight Activities Officer
FC .... Flight Controller
FCO .... Flight Control Office
FCSM .... Flight Combustion Stability Monitor
FD .... Flight Director
FDAI .... Flight Director Attitude Indicator
FDK .... Forced Display Module
FDO .... Flight Dynamics Officer
FET .... Flight Evaluation Team
FEWG .... Flight Evaluation Working Group
FGR .... Flight Geometry Reserves
FLSC .... Flexible Linear Shaped Charge
FLTR .... Filter
FM .... Frequency Modulation
FFPR .... Flight Performance Reserves
FRR .... Flight Readiness Review
FRT .... Flight Readiness Test
FT .... Foot or Feet
FTS .... Flight Termination System

G

G .... Acceleration of Gravity
GAL .... Gallon(s)
GBI ...... DOD-ETR MSFN Station, Grand Bahama Island
GBM .... NASA MSFN Station, Grand Bahama Island
GCC .... Ground Control Computer
GDC .... Gyro Display Coupler
GDS .... NASA MSFN Deep Space Station, Goldstone
California
GDSX .... MSFN Wing of JPL Station at Goldstone
GET .... Ground Elapsed Time
GG .... Gas Generator
GH₂ .... Gaseous Hydrogen
GHz .... Gigihertz (One Billion Hertz)
GN₂ .... Gaseous Nitrogen
G&C .... Guidance and Control
G&N .... Guidance and Navigation
GNC .... Guidance, Navigation and Control
GND .... Guidance, Navigation and Digital Systems
Engineer (Booster Systems)
GOX .... Gaseous Oxygen
GPF .... Grains per Foot
GPI .... Gimbal Position Indicator
gpm .... Gallons per Minute
GRR .... Guidance Reference Release
GSE .... Ground Support Equipment
GSE-ECU .... GSE-Environmental
Control Unit
GSFC .... Goddard Space Flight Center
GTK .... DOD-ETR MSFN Station, Grand Turk Island
GUID .... Guidance
GUIDO .... Guidance Officer

G-V .... Gravity versus Velocity
GWM .... NASA MSFN Station, Guam

H

H₂ .... Hydrogen
Ḣ .... Altitude Rate of Change
HAW .... NASA MSFN Station, Hawaii
HD .... Highly Desirable Mission Rule Item
He .... Helium
HF .... High Frequency (3-30 MHz)
HOSC .... Huntsville Operations Support Center
HSD .... High-Speed Data
HSK... WRE MSFN Deep-Space Station, Honeysuckle Creek,
Australia
HSKX ....MSFN Wing of JPL Station at Honeysuckle Creek
HTV .... DOD-WTR MSFN Station, USNS Huntsville
Hz .... Hertz (one cycle per second)

I

IA .... Input Axis
IBM .... International Business Machines Corporation
ICD .... Interface Control Document
ICO .... Inboard Cutoff
IECO .... Inboard Engine Cutoff
IF .... Intermediate Frequency
IFV .... Igniter Fuel Valve
IGM .... Iterative Guidance Mode
IMP .... Impulse
IMU .... Inertial Measurement Unit
IMV .... Ignition Monitor Valve
in .... Inch
INTELSAT .... Communications Satellite
IO .... Industrial Operations
IOS .... Indian Ocean Ships
IP .... Impact Predictor (Located at Kennedy Space
Center)
IRIG .... Inter-Range Instrumentation Group
IU .... Instrument Unit
IVT .... Intravehicular Transfer

J

JPL .... Jet Propulsion Laboratory

K

kbps .... Kilobits per Second
kHz .... Kilohertz (One Thousand Hertz)
KM .... Kilometer
KOH .... Potassium Hydroxide
KSC .... Kennedy Space Center

L

lb(s) .... Pound(s)
LCC .... Launch Control Center
LCR .... LIEF Control Room
LE .... Launch Escape
LEM .... Lunar Excursion Module
LES .... Launch Escape System
LET .... Launch Escape Tower
LH₂ .... Liquid Hydrogen
LIEF .... Launch Information Exchange Facility
LLM .... Lunar Landing Mission
LM .... Lunar Module
LO .... Liftoff

LOR . . . . Lunar Orbital Rendezvous
LOS . . . . Loss of Signal
lox . . . . Liquid Oxygen
LSC . . . . Linear Shaped Charge
LSD . . . . Low-Speed Data
LTA-B . . . . Lunar Test Article, Model B
LUT . . . . Launch-Umbilical Tower
LV . . . . Launch Vehicle
LVDA . . . . Launch Vehicle Data Adapter
LVDC . . . . Launch Vehicle Digital Computer

M

M . . . . Mass or Mandatory Mission Rule Item
MA . . . . Apollo Program Office (Symbol)
MAD . . . . . NASA MSFN Deep Space Station, Madrid, Spain
MADX . . . . MSFN Wing of JPL Station at Madrid
MAP . . . . Message Acceptance Pulse
MAX . . . . Maximum
MCC . . . . Mission Control Center
MCP . . . . Mission Control Programmer
MCR . . . . Main Conference Room
MD . . . . Mission Director
MDC . . . . Main Display Console
MDC . . . . McDonnell-Douglas Corporation
MDF . . . . Mild Detonating Fuse
MED . . . . Medium
MER . . . . DOD-WTR MSFN Station, USNS Mercury
MESC . . . . Master Event Sequence Controller
MFCO . . . . Manual Fuel Cutoff
MHz . . . . Megahertz (One Million Hertz)
MIL . . . . NASA MSFN Station, Merritt Island, Florida
MILA . . . . Merritt Island Launch Area
MIN . . . . Minimum
Min . . . . Minute
ML . . . . Mobile Launcher
MLA . . . . . DOD-ETR MSFN Station, Merritt Island, Florida
MMH . . . . Monomethyl Hydrazine
MNA . . . . Main Bus A (Spacecraft)
MNB . . . . Main Bus B (Spacecraft)
M&O . . . . Maintenance and Operations
MOC . . . . Mission Operations Computer
MOCR . . . . Mission Operations Control Room
MOD . . . . Modification or Model
MR . . . . Mixture Ratio
MSC . . . . Manned Spacecraft Center
m/sec or ms . . . . Millisecond (1/1000 Second)
MSFC . . . . Marshall Space Flight Center
MSFN . . . . Manned Space Flight Network
MSFNOC . Manned Space Flight Network Operations Center
MSK . . . . Manual Select Keyboard
MSR . . . . Mission Support Room
MSS . . . . Mobile Service Structure
MTF . . . . Mississippi Test Facility
MTVC . . . . Manual Thrust Vector Control
MUX . . . . Multiplexer

N

N/A . . . . Not Applicable
NASA . . . . . National Aeronautics and Space Administration
NASCOM . . . . NASA Communications Network
NMI . . . . Nautical Mile
No . . . . Number
NORAD . . . . North American Air Defense Command (DOD)
NPSH . . . . Net Positive Suction Head

O

OA . . . . Output Axis
OAT . . . . Overall Acceptance Test
OBECO . . . . Outboard Engine Cutoff
OCO . . . . Outboard Cutoff
ODOP . . . . Offset Doppler
OECO . . . . Outboard Engine Cutoff
O₂ . . . . Oxygen
O&P . . . . Operations and Procedures
OSC . . . . Oscillator
OSR . . . . Operations Support Room
OTV . . . . Operational Television System
OX . . . . Oxidizer

P

P . . . . Pitch
PAFB . . . . Patrick Air Force Base, Florida
PAM . . . . Pulse Amplitude Modulation
PAT . . . . . DOD-ETR MSFN Station, Patrick AFB, Florida
PBI . . . . Push-Button Indicator
PC . . . . Pitch Control
PCM . . . . Pulse Code Modulation
pct . . . . Percent
PDFRR . . . . . . Program Director's Flight Readiness Review
PDR . . . . Preliminary Design Review
PDS . . . . Propellant Dispersion System
PEA . . . . Platform Electronics Assembly
PETN . . . . Pentaerythrite Tetranitrate
PFR . . . . Preflight Review
PGNCS . . . Primary Guidance Navigation and Control System
PLSS . . . . Portable Life Support System
POGO . . . . Undesirable Launch Vehicle Longitudinal
Oscillations
POI . . . . Parking Orbit Insertion
POS . . . . Pacific Ocean Ships
POS . . . . Position
POT . . . . Potentiometer
pps . . . . Pulses per Second
PRE . . . . . DOD-ETR MSFN Station, Pretoria, So. Africa
PRF . . . . Pulse Repetition Frequency
PRN . . . . Psuedo-Random Noise
PRPLNT . . . . Propellant
psi . . . . Pounds per Square Inch
psia . . . . Pounds per Square Inch Absolute
psid . . . . Pounds per Square Inch Differential
psig . . . . Pounds per Square Inch Gauge
PTCR . . . . Pad Terminal Connection Room
PTCS . . . . Propellant Tanking Computer System
PTL . . . . Prepare to Launch
PU . . . . Propellant Utilization

Q

QLDS . . . . Quick Look Data Station

R

R . . . . Roll
RACS . . . . Remote Automatic Calibration System
RASM . . . . Remote Analog Submultiplexer
RCR . . . . Recovery Control Room
RCS . . . . Reaction Control System
RDM . . . . Remote Digital Multiplexer
RDX . . . . Cyclotrimethylene-trinitramine
R&DO . . . . Research and Development Operations

RDSM	Remote Digital Submultiplexer
RED	DOD-WTR MSFN Station, USNS Redstone
RETRO	Retrofire Officer
RF	Radio Frequency
RFL	Full Lift Landing Range
RHL	Half Lift Landing Range
RI	Radio Interference
RIS	Range Instrumentation Ship
RMS	Root Mean Square
RNG	Range
RP-1	Rocket Propellant
R&QA	Reliability & Quality Assurance
R&R	Receive and Record
RS	Relay Station
RSCR	Range Safety Command Receiver
RSDP	Remote Site Data Processor
RSO	Range Safety Officer
RSS	Root Sum Square
RTC	Real Time Command
RTCC	Real-Time Computer Complex
R/T CMD	Real-Time Command Controller

S

S&A	Safety and Arming
SACTO	Sacramento Test Operations
SC	Spacecraft
SCAMA	Switching, Conferencing, and Monitoring Arrangement
scfm	Standard Cubic Feet per Minute
SCO	Subcarrier Oscillator
SCS	Stability Control System
SDF	System Development Facility
sec	Second
SECO	Single Engine Cutoff
SEP	Separation
SIT	Systems Interface Test
SLA	Spacecraft-LM Adapter
SLV	Saturn Launch Vehicle
SM	Service Module
SMEK	Summary Message Enable Keyboard
S/N	Serial Number
SPS	Service Propulsion System
SRA	Spin Reference Axis
SRM	Status Report Module
SRO	Superintendent of Range Operations
s/s	Samples per Second
SSB	Single Sideband
SSR	Staff Support Room
ST-124-M3	Saturn V Stable Platform
STDV	Start Tank Discharge Valve
SV	Space Vehicle
SW	Switch(ing)
SYNC	Synchronize

T

T (With Subscript)	Time Base
TAN	NASA MSFN Station, Tananarive, Malagasy
TBD	To Be Determined
TCS	Thermal Conditioning System

TD&E	Transposition, Docking and Ejection
TEL IV	AFETR Telemetry Station at Cape Kennedy
Teltrac	Telemetry Tracking
TEX	NASA MSFC Station, Corpus Christi, Texas
TLI	Translunar Injection
TM	Telemetry
TPF	Terminal Phase Finalization
TPF	Terminal Phase Finalization
TPI	Terminal Phase Initiation
TSM	Tail Service Mast
TTY	Teletype
TV	Television
TVC	Thrust Vector Control
TWR JETT	Tower Jettison

U

UDL	Up-Data-Link
UHF	Ultra High Frequency (300-3000 MHz)
USB	Unified S-Band
USBS	Unified S-Band Station
USNS	United States Navy Ship

V

V	Velocity
V	Voice
v	Volts
v-a	Volt-ampere
VAB	Vehicle Assembly Building
VAN	DOD-ETR MSFN Station, USNS Vanguard
V/D	Voice and Data
vdc	Volts, Direct Current
VHF	Very High Frequency (30-300 MHz)
vswr	Voltage Standing Wave Ratio

W

WBD	Wide Band Data
WRE	Weapons Research Establishment, Australian Department of Supply
WSMR	White Sands Missile Range
WTR	Western Test Range

X-Z

XLUNAR	Translunar
XTAL	Crystal
Y	Yaw

SYMBOLS

β	Thrust Vector Angular Deflection
ΔP	Differential Pressure
ΔV	Velocity Increment
$\Delta \rho$	Atmospheric Density Increment
μ sec	Microsecond
ϕ	Vehicle Attitude
ϕ'	Vehicle Attitude Angular Change Rate
ψ	Attitude Error Command
χ	Desired Attitude

APPENDIX B

BIBLIOGRAPHY

Abort and Malfunctioned Flight Analysis, Saturn V Launch Vehicle, AS-505; April 2, 1969; Boeing Report D5-15795-5.

Apollo Abort Summary Document, Mission G, (AS-506/107), Final; May 21, 1969; MSC Internal Note MSC-CF-P-69-30.

Apollo Design Certification Review; April 21, 1966; Apollo Program Directive No. 7.

Apollo Flight Mission Assignments; July 11, 1969; SE 010-000-1.

Apollo Flight Readiness Reviews, Part I; November 8, 1965; Apollo Program Directive No. 8.

Apollo Mission Rules; April 12, 1967; NASA Management Instruction 8020.9.

Apollo Operations Handbook, Block II Spacecraft, Volume 1, Spacecraft Description; September 27, 1968; SM2A-03-Block II -(1).

Apollo Operations Handbook, Command and Service Module, Volume 2, Operational Procedures; December 2, 1968; SM2A-03-SC106-(2).

Apollo Program Directive No. 4K; July 10, 1969.

Apollo/Saturn V Facility Description; October 1, 1966; K-V-012.

Apollo Test Requirements; March 1967; NHB 8080.1.

Astrionics System Handbook, Saturn Launch Vehicles; November 1, 1968; MSFC No. IV-4-401-1.

Automatic Terminal Countdown Sequence Interlocks SA-502; February 16, 1968; 40M50547.

Battery Assembly; May 19, 1965; 40M30780.

Battery, S-IVB Aft Bus No. 2; Rev. L, July 22, 1968; 1A68317.

Battery, S-IVB Forward Bus No. 1 and Aft Bus No. 1; Rev. N; August 2, 1968; 1A59741.

Battery, S-IVB Forward Bus No. 2; Rev. M; July 19, 1968; 1A68316.

Battery Technical Manual; February 15, 1968; IBM 66-966-0024.

C-Band Radar Transponder Set SST-135C; March 1, 1967; IBM 66-966-0008.

Clearance Envelope, LEM/S-IVB/IU, Physical; Rev. B; Not dated; 13M50123.

Command Decoder Technical Manual; March 1, 1967; IBM 66-966-0019.

Components Clearance Envelope S-IVB to IU; March 15; 1968; 13M50117.

Composite Mechanical Schematic; Rev. K; February 1, 1969; 10M30531.

Countdown Hold Limits and Vehicle Turnaround Operations, AS-504; February 20, 1969; Boeing Letter Report 5-6171-H-130.

Coupled Loads Analysis from Ignition to S-IC/S-II Separation, AS-507; August 15, 1969; Boeing Letter Report 5-9310-H-070.

Design Certification Review Report S-IC Stage, AS-503, Volume 4; September 15, 1968; Not numbered.

Design Certification Review Report, S-II-3 Stage; June 17, 1968; SD68-125.

Design Certification Review Report, S-IVB Stage, AS-503, Volume 6; August 23, 1968; DAC-56644.

Electrical Interface, Launch Vehicle ESE/Propellant Systems, Saturn V Vehicles/LC-39; Rev. B; September 10, 1968; ICD 40M05486B.

Electrical Interface, Saturn V Launch Vehicle and Apollo Block II CSM; January 19, 1968; 40M37521.

Electrical Interface, S-IC/S-II, Saturn V Vehicle; Rev. C; May 29, 1968; 40M30596.

Electrical Interface, S-II/S-IVB, Saturn V Vehicle; Rev. C; June 26, 1968; 40M30593.

Electrical Interface, S-IVB/Instrument Unit, Saturn V Vehicle; Rev E; September 30, 1968; 40M30597.

Emergency Detection and Procedures (Draft Copy) Received from MSFC November 7, 1967; Not numbered.

Emergency Detection and Procedures, Addendum, (Draft copy); Received from MSFC November 17, 1967; Not numbered.

Emergency Detection System Description, Apollo-Saturn V; July 1, 1967; IBM 66-966-0022.

F-1 Rocket Engine Data Manual; Chg. 6; August 14, 1968; R-3896-1.

F-1 Rocket Engine Operating Instructions; Chg. 4; August 6, 1969; R-3896-11.

Flight Control Data Requirements, Saturn Launch Vehicle, Standard; August 1, 1967; Not numbered.

Flight Control, Introduction to; June 16, 1967 SCD No 1 01.

Flight Control System, Saturn S-II; February 1, 1965; SID 62-140.

Flight Dynamics Analysis, Saturn V Launch Vehicle, AS-507; July 24, 1969; Boeing Report D5-15509(F)-7.

Flight Manual, Saturn V, SA-506; June 10, 1969; MSFC-MAN-506.

Flight Manual, Saturn V, SA-507; August 15, 1969; MSFC-MAN-507.

Flight Operations Planning and Preparation for Manned Orbital Missions; (John H. Boynton and Christopher C. Kraft, Jr.) AIAA Paper No. 66-904.

Flight Performance Handbook for Powered Flight Operations; (Space Technology Laboratories, Inc.) Not dated; Not numbered.

Flight Sequence (H-Mission), Saturn V/SA-507; March 11, 1969; 10M30637.

Flight Sequence Program, Definition of, Saturn SA-507; August 4, 1969; 40M33627.

Fluids Requirements, Saturn V Vehicle Instrument Unit; ICD, Rev. C; June 13, 1968; 13M50099.

Fluids Requirements, Saturn V Vehicle S-IC Stage; ICD, Rev. F; March 24, 1969; 13M50096F.

Fluids Requirements, Saturn V Launch Vehicle, S-II Stage; ICD, Rev. H; April 30, 1969; 13M50097H.

Fluids Requirements, Saturn V Launch Vehicle, S-IVB Stage; ICD, Rev. E; February 19, 1969; 13M50098E.

Frequency Plan, Saturn Apollo, Instrumentation and Communications; Rev. C; May 2, 1968; 13M60003.

Fuel Tank Pressurization System S-IVB, Advanced Functional Schematic; Rev. L; November 21, 1967; 1B62829.

Ground Safety Plan, Apollo/Saturn IB, April 1, 1966; K-1B-023.

Ground Support Equipment Fact Booklet, Saturn V Launch Vehicle; Chg. March 15, 1968; MSFC-MAN-100.

Hazardous Gas Detection System; April 15, 1968; MSFC-MAN-234.

Inboard Profile, Saturn V; Rev. M; February 25, 1966; 10M03369.

Instrumentation Subsystem Report; July 1, 1967; SID 62-136.

Instrument Unit Assembly; Rev. E, F, and H; 10Z22501.

Instrument Unit Environmental Control System Description, Saturn IB/V; February 1, 1968; IBM 66-966-0009.

Instrument Unit-Saturn V, Functional Interlock Requirement; Rev. A; May 16, 1966; 10M30554.

Instrument Unit System Description and Component Data, Saturn IB/V; Chg. July 1, 1968; IBM 66-966-0006.

J-2 Rocket Engine Data Manual; Chg. 6; September 18, 1969; R-3825-1.

J-2 Rocket Engine Operating Instructions Manual; Chg. 5; July 31, 1969; R-3825-1B.

Launch Vehicle Operations for Support of Space Vehicle Countdown Demonstration Test and Launch Countdown, Vol. II of III; April 2, 1969; KSC V-20060.

Lox Tank Pressurization System, S-IVB, Advanced Functional Schematic; Rev. D; November 11, 1967; 1B62828.

Mission Implementation Plan, H-Series Missions, Apollo 12/AS-507, 13/AS-508, 14/AS-509 and 15/AS-510; August 29, 1969; PM-SAT-8010.5B.

Mission Summary Report, AS-507, H-1 Mission; August 29, 1969; Boeing Letter Report 5-9400-H-360.

Mobile Launcher, Vehicle Holddown Arms and Service Platforms, Apollo Saturn V, Operations and Maintenance; March 25, 1969; TM 487-MD.

Navigation, Guidance, and Control System Description; November 1, 1966; IBM 66-966-0003.

Network Operations Document All Flights, Goddard Space Flight Center; June 19, 1969; Not numbered.

Operational Flight Trajectory, Saturn V AS-507 Launch Vehicle, H-1 Mission, November Launch Month; August 12, 1969; Boeing Report D5-15551(F)-7.

Operational Mass Characteristics, Saturn V, AS-507; June 30, 1969; D5-15512-7.

Ordnance Systems, Saturn V Launch Vehicle; August 1, 1969; SE 005-003-2H.

Performance Prediction, Propulsion System, SA-507; Vol. I (S-IC), Vol. II (S-II) and Vol. III (S-IVB); May 29, 1969; Boeing Report D5-15530-7.

Propulsion System Performance Prediction, S-IC, SA-507; August, 1969; Boeing Report D5-13699-7.

Protective Hood and Tower Installation, Tail Service Mast; January 26, 1968; 76K04274.

Range Safety; September 10, 1965; AFETRR 127-9.

Range Safety Manual; September 10, 1965; AFETRM 127-1.

S-IVB-503N Stage Flight Information and Operations Report-C Prime Mission, Volume I; November 1968; DAC-56675A.

S-IVB-506N Stage Flight Information and Operations Report-G Mission; Revised June, 1969; DAC-56773A.

Schedules and Status, Apollo/Saturn V; Updated Weekly; PM-SAT-PS.

BIBLIOGRAPHY

Secure Range Safety Command System for Saturn; November 9, 1964; NASA TM X-53162.

Sequence and Flow of Hardware Development and Key Inspection, Review , and Certification Checkpoints; August 30, 1966; Apollo Program Directive No. 6A.

Sequence of Events, Prelaunch and Flight, AS-506; February 20, 1969; 10M30586.

Slide Wire Egress System, LC-39; March 15, 1969; TM-4-214-39.

Specification Baseline, Saturn V Program (SA-504 through SA-515); September 23, 1966 (Revised October 29, 1968); RS-02W-1000004A.

Specification, Detail, Contract End Item, Saturn V/S-IC (Prime Equipment), SA-503 and subsequent, Performance/Design and Product Configuration Requirements; September 5, 1969; CP02S00001103D.

Specification, Detail, Contract End Item, Saturn V/S-II (Prime Equipment), SA-503 and subsequent, Performance/Design and Qualification Requirements; January 9, 1969; CP621M0014A.

Specification, Detail, Contract End Item, Saturn V/S-IVB (Prime Equipment), SA-507 through SA-515, Performance/Design and Qualification Requirements; January 1, 1969; CP209000B.

Structural Design Assurance, Saturn V Launch Vehicle, AS-504; January 2, 1969; Boeing Report D5-15568-4.

System Description Manual, S-IC Stage, AS-503 thru AS-505; October 1, 1968; MSFC-MAN-0040-3, 4, 5.

Telemetry System Manual, Saturn V, Vol. 3; March 10, 1969; 50M71535-3.

Timer, 30-Second, Multiple Engine Cutoff Enable; May 22, 1968; Rev. C; 7915671-1.

This Space Intentionally Left Blank

ALPHABETICAL INDEX

index

Printed in September 2023
by Rotomail Italia S.p.A., Vignate (MI) - Italy